I0468646

Outside Plant Fiber Optics Construction Guide

Joe Botha

The Fiber Optic Association, Inc.

The professional society of fiber optics

The FOA Outside Plant
Fiber Optics Construction Guide

Joe Botha

The Fiber Optic Association, Inc.

1119 S Mission Road #355, Fallbrook, CA 92028
Telephone: 1-760-451-3655   Fax: 1-781-207-2421
Email: info@thefoa.org http://www.thefoa.org

ISBN 1523925825

The FOA Outside Plant Fiber Optics Construction Guide

Preface

The Fiber Optic Association, Inc., the nonprofit professional society of fiber optics, has become one of the principal sources of technical information, training curriculum and certifications for the fiber optic industry. The FOA printed textbooks, Online Guide (www.foaguide.org) and Fiber U, the FOA's free online trianing website (www.fiberu.org) are major sources of technical information for the fiber optic community.

This textbook is a guide to outside plant fiber optic construction, basically the process of installing the fiber optic cable plant including the work necessary before the fiber optic techs begin splicing, terminating and testing the cable plant. This book was written by Joe Botha of Triple Play Fibre Optic Solutions in South Africa as a textbook for classes he teaches on construction. Joe, an FOA Master Instructor, created the course to fill a need for training OSP construction crews. The book covers topics which are rare in textbooks, practical solutions to designing and installing the fiber optic cable plant. It is an extremely valuable reference book for all owners, designers, supervisors and installers of fiber optic OSP networks.

If you have feedback on the book, feel free to email comments or questions to the FOA at info@foa.org.

A note of appreciation

The FOA wishes to thank Master instructor Joe Botha of Triple Play Fibre Optic Solutions for his work in creating this book and sharing it with the fiber optic community. We also wish to that the other FOA members who have reviewed this book and convinced FOA to publish it for everyone's use.

Contents

Why Fibre?

Since the introduction of fibre in the 1970s, optical fibres have revolutionised communications, transmitting more information over greater distances than could ever be achieved in copper wires.

We live on the continent (Africa) that gave birth to the concept of the 'Digital Divide'. But this being said, fast forward a few years, and today, the uncompromising amount of bandwidth we demand from Internet services is nearly insatiable. Why then do we need all that bandwidth? The past 10 years have seen a tremendous increase in Internet bandwidth requirements ("Moore's law" springs to mind), driven by high-capacity business data services, increasingly powerful 3G/4G wireless smartphones and video-intensive Websites such as YouTube, Netflix, Hulu, Amazon's Prime and locally, ShowMax. Hands-up who doesn't watch multiple episodes of a good TV series in one sitting. Binge-watching has already fuelled the growth of streaming service Netflix, which now accounts for 37% of all North American Internet traffic during prime time! Gloomy circumstances for those without FTTH. And, of course, lets for forget about the Periscope and Meerkat live-streaming video apps.

Buoyed by the surge in demand for digital data, local telecommunication companies are looking to reinvent themselves to support the need for faster data rates and smarter network architectures, potentially hamstringed by the predicament of having to handle unpredictable and fast-changing traffic patterns. Optical fibre technology provides a higher capacity data transfer at extremely high speeds, enabling the facilitation of video content on networks. In this context, community or service providers can now supply a wide range of services and applications, such as High Definition TV (HDTV), Video on Demand (VoD) and high-speed data and on top of this, provide for the basic fundamentals of voice connectivity.

Switching a laser beam on and off at high speeds is how we communicate now over fibre. A laser flashes on and off in a sequence that represents the information being sent. The quicker the lasers can flash on and off, the quicker the information can be transferred. The speed of transfer is known as "bit-rate" and is usually talked about in terms of bits per second - bps or bit/s (which effectively is the number of flashes possible per second). Modern networks typically transmit at a minimum of 10 Gigabits per second from a single laser, which amounts to 10 billion flashes per second.

And goodness me, Alcatel-Lucent recently (Jul 16, 2013) transmitted as much as 31 Terabits per second over 7200 km on a single fibre i.e. 155 x 200-Gbps channels. Truly astonishing, hey? This link featured amplifiers every 100 km and as far as I know, the highest sub-sea capacity ever transmitted on a single optical fibre.

Fibre Advantages - a brief summary

- It has exceptional bandwidth and is virtually "future proof"
- It has the ability to carry many signals concurrently
- It is immune to electromagnetic interference and has no electromagnetic emissions
- It does not corrode like copper based cabling does and is resistant to eavesdropping
- It has the capability to operate in conjunction with any current, or proposed, LAN/WAN standard
- It is light weight and easy to handle and it's pulling strength is higher than that of copper cables
- ...and, on it goes

What happens to Johannesburg in spring? It sprouts leaves and changes from somewhat bare and dry conditions in winter, to lush vegetation in spring. Sadly, radio waves struggle to penetrate vegetation. So along with spring, come slower download speeds, poor signals and of course, disgruntled customers. The solution? More fibre, of course!

Joe Botha
Triple Play Fiber Optic Solutions

Fibre Geometric Parameters

Fibre can either be single-mode (SM) or multimode (MM). Fibre sizes are expressed by using two numbers e.g. 9/125. The first number refers to the core size in microns and the second number refers to the core and cladding size combined in microns. It is impossible to differentiate between SM and MM fibre with the naked eye. There is no difference in the outward appearances; both are 125 microns in size - only the core size differs.

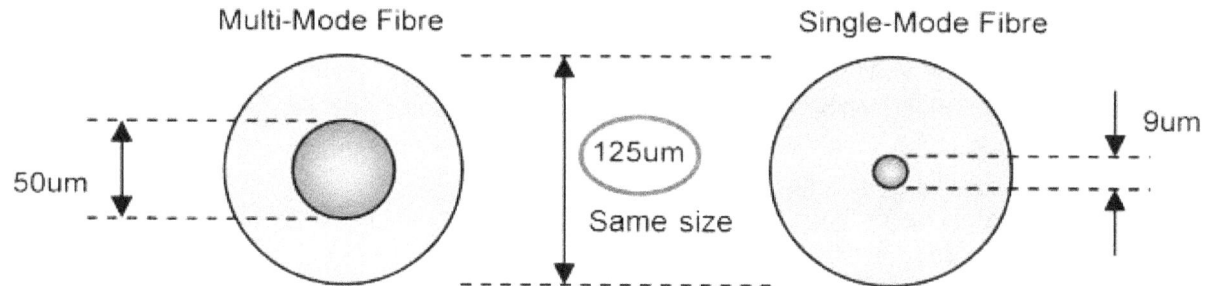

Fibre Construction:

Core: The optical core is the light-carrying element at the centre and is usually made up of a combination of silica and germanium.

Cladding: Cannot be removed! The cladding surrounding the core is made of pure silica and has a slightly lower index of refraction (i.e. less dense) than the core. This lower refractive index causes the light in the core to reflect when encountering the cladding and remain trapped within the core.

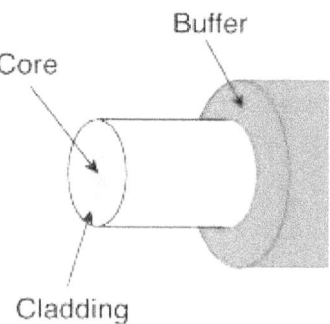

Buffer coating: This is removed during stripping for splicing or connectorization and acts as a shock absorber to protect the core and cladding from damage.

Cable Jackets

Polyethylene (PE) is the material of choice for use as an Outside Plant (OSP) cable jacket. The performance of raw PE can degrade rapidly through exposure to Sunlight. For this purpose, Carbon Black is combined with the PE and is used to absorb the UV light and subsequently dissipates. Jacket colours other than black are used for reasons of enhancing identification.

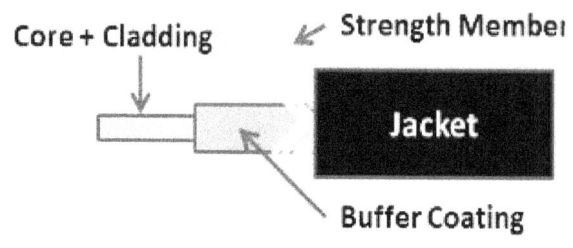

Cable jackets shall be marked with manufacturer's name, month and year of manufacture, sequential meter markings, fibre type, the number of fibres, along with a telecommunications handset symbol. Cables without these markings will not pass inspections and should not be installed.

Strength members

Aramid fibres (Kevlar - a very strong, very light, synthetic compound developed by DuPont – is used when a cable is pulled into a duct, with the tension being applied to the Kevlar. The Kevlar is used as a draw string to pull the cable into the duct so as not to put stress on the fibres.

The term is also used for the fibreglass or steel rod in some cables used to stiffen it. Impact resistance, flexing and bending are other mechanical factors affecting the choice of strength members.

Moisture/Water-blocking

In a loose tube cable design, a filling compound, water swell-able yarns or gel (a soft, gooey, substance) are commonly incorporated in the cable. This minimizes the chance of water or moisture penetrating the length of the tube in the event that the tube is damaged. When water freezes it expands by approximately 9% - therefore, water in a cable can cause repeated freeze and thaw cycles.

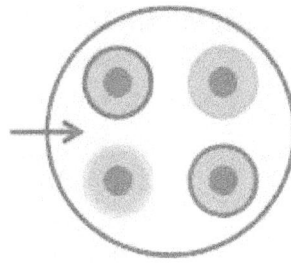

Micro Technology

Since SM fibre was first introduced in the early 1980s, not too much has changed in its basic geometric parameters. The SM core size has remained somewhere between 8 and 10 μm - the core / cladding diameter has remained at 125 μm. The outside plant (OSP) coating is now either 200 or 250 μm. Most notably, standardising these dimensions has greatly improved interoperability and consistency across optical networks.

A new-ish component to the system is microducts made of high-density polyethylene (HDPE) material. They are typically installed as bundles in larger ducts. It's generally accepted, that deploying microducts is an economically sensible option, helping to avoid the costs of subsequent civil works.

Microcables offers a great deal more density. Only a few years ago, a cable diameter of ± 12 mm was required for a 48-fiber cable design. Today, cables with only an 8 mm diameter has a capacity of 144-fibres. This is achieved by using 200μm coated fibres. The 200 μm coated fibre's cross sectional area is ± 46% smaller than that of the conventional 250 μm coated fibre.

Micro cables are designed for jetting, with high-density polyethylene (HDPE) outer sheaths in order to minimize friction with the inner surface of micro ducts. Perhaps more significantly, these cables also have optimal stiffness properties to help prevent buckling and easily negotiate modest changes in direction of the micro duct along the jetting route.

The central tube micro cable design also provides the highest fibre density, yielding a relatively small cable OD. The individual fibres are bundled into groups of twelve within the cable's central tube, and the bundles are easily identifiable with coloured binders in accordance with EIA/TIA-598B, "Optical Fibre Cable Colour Coding".

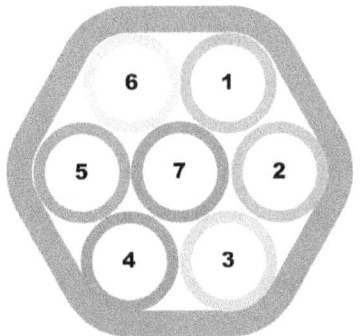

Microduct Colour Code
1. Blue
2. Orange
3. Green
4. Brown
5. Slate
6. White
7. Red

NB, colour codes may differ from supplier-to-supplier.

EIA/TIA-598B, "Optical Fiber Cable Colour Coding"

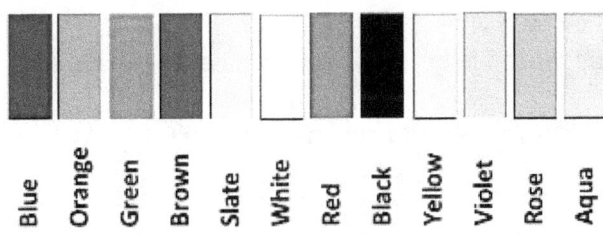

Blue   Orange   Green   Brown   Slate   White   Red   Black   Yellow   Violet   Rose   Aqua

Environmental Considerations
It is the Contractor and/or Client's responsibility to prepare a site-specific Environmental Management Plan (EMP). And, as ever, each site is unique and therefore the environmental issues to be considered will vary from site to site.

Environmental Control Officer (ECO)
1   A good place to start with, is the appointment of an Environmental Control Officer (ECO).
2   In order for all to sleep easily at night, without having to furrow the brow much with care, the ECO shall:
    - Prior to work commencing, the contractor shall unravel the requirements of the EMP to team members, to ensure understanding and conformity
    - Visit the site weekly
    - Review and approve all areas that have been rehabilitated by the Contractor
    - Keep a record of findings
    - Attend all project meetings
    - Produce a monthly report, for the client, with commentary on compliance and/or non-compliance
    - Keep an Incident Log of non-compliance
    - Maintain a record of complaints from the public and communicate this to the client
    - Where necessary, issue a non-compliance report to the Contractor
    - Where serious environmental infringements have occurred, introduce a Temporary Work Stoppage
    - Liaise with the appointed Occupational Health and Safety Officer of the client

The Contractor's Environmental Management Plan (EMP) must include:
In what follows, is a brief summary of what an EMP might contain, but is not limited to:
1. Statement of Commitment
2. List of activity-specific environmental issues related to your site and their likely impact
3. Incorporating the above, write a series of simple work instructions to ensure compliance
4. Determine the actions required to manage each work instruction
5. A list of tangible contingency and mitigating actions to be implemented if required
6. Provide training to staff and create awareness
7. An organisational chart setting out respective roles and responsibilities
8. Monitoring and Reporting

Contractor Responsibilities (not limited to):
1   Be familiar with and comply with the procedures contained in the EMP.
2   Ensure that all personnel are trained, qualified and experienced enough, to undertake their work in an environmentally responsible manner.
3   Create an awareness of the Environmental work requirements and the need for them amongst the workforce.
4   Procedural briefings to be given before personnel carry out key activities for the first time.
5   Ensure that personnel who have formal responsibilities under this plan, are trained in the requirements of this EMP.
6   Undertake daily site inspections to monitor environmental performance and compliance.
7   Immediately notify the ECO in the event of infringements.
8   Notify the ECO and in advance of any activity he has reason to believe that may potentially have an adverse environmental impact.
9   It is desirable for Environmental matters to be included as a standard agenda item at all project meetings.

A few Activity-specific Guideline Examples:

In what follows, are a few activity based examples an EMP must contain, but is not limited to:

Water
1 Storm water must be contained in the storm water system to avert flooding.
2 Measures must be implemented to distribute storm water as evenly as possible to fend-off soil erosion.
3 Material from any stockpile must not be allowed to spill or be washed into a gutter or drain.
4 The execution of any work shall not block and subsequently unsettle the existing overland water flow or the existing system of drains.
5 No person may, without prior written permission release water onto a public road.
6 No person may, without prior written permission raise the water level of a river, stream or dam which can spill-over onto a public road.
7 Any water that is present in a trench shall be pumped out before backfilling.
8 Water shall be pumped into the storm water system and never into a sewer manhole.
9 No work shall be carried out within 32m of any natural water source without permission from the ECO.

Dust / Air Pollution
1 Where necessary, issue workers with washable dust masks for protection against dust inhalation.
2 Dust shall be controlled onsite, especially when windy.
3 Minimize or even cease activity during periods of high wind.
4 Dampen surfaces to prevent dust from becoming airborne.
5 Cover materials being transported with plastic sheeting or a tarpaulin to prevent them from flying off the vehicle. Dampening of the transported material may also be necessary.
6 Cover onsite stockpiles with plastic sheeting or tarpaulins during high wind.
7 Regular maintenance of generators, compressors, etc., is essential for controlling exhaust emissions.

Noise Pollution
1 Contractors must abide by the National Noise laws.
2 Develop a noise mitigation plan before starting with construction.
3 The level of noise and the duration thereof must be agreed upon and monitored.
4 Examples of noise: Jack hammers, concrete saws, bulldozers, trucks, generators, compressors, pneumatic tools, power tools, etc.
5 When talking to someone 1m away and you have to shout to make yourself heard, then noise levels are definitely too high.
6 Hand-held sound meters are lightweight, easy to operate and relatively inexpensive.

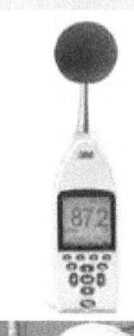

7 Hearing protection shall be worn at all times when noise levels are suspected of equalling or exceeding 85 dBA.
8 Noise PPE:
   - Use disposable earplugs only once
   - Keep reusable earplugs clean
   - Earmuffs must be a good fit

9 Where possible, restrict construction work to weekdays and limit work hours from 08:00 to 17:00.
   Should an extension of the work hours be required, the adjacent property owners shall be informed
10 in writing 2-days in advance of any proposed overtime activities.
11 Whenever practical, noise levels identified as exceeding 85 dBA, Noise MUST be reduced by using mufflers, barriers, etc. or the following actions must be implemented once the source of the noise has been ascertained:

- Replacement or adjustment of the worn or loose parts
- Balancing unbalanced equipment
- Lubrication of the moving parts
- Use of properly shaped and sharpened cutting tools

Trees

1 Trees shall not be cut or trimmed unless consent is obtained in writing from the owner or the relevant authority.
2 Cutting shall be confined to what is absolutely necessary.
3 Tree roots exposed in the way of the trenching shall not be cut unless absolutely unavoidable.

Archaeology and Cultural Heritage

1 Even with no designated sites of archaeological sensitivity being identified along a route, the Contractor will be required to have measures in place to deal with potential finds protected by the Natural Heritage Resources Act, Act 25 of 1999.
2 Construction in the vicinity of a finding must be stopped and under no circumstance may any artefacts be disturbed or removed from the site.
3 An archaeologist can be called to the site for inspection and the South African Heritage Agency must be advised.

Environmentally Sensitive Areas (ESA)

1 An ESA is a type of designation for an agricultural area which needs special protection because of its landscape, wildlife or historical value.
2 Such designated areas shall be dubbed "no go" areas and access to, or work in such areas, shall be carefully controlled by the ECO.

Concrete / Cement

1 Concrete shall not be mixed directly on the ground.
2 All visible remains of excess concrete shall be removed and disposed at an approved disposal site.
3 Cement shall be stored in a dry place, protected from rain and raised off the floor.

Bird Nests

1 Bird nests cannot be moved, unless consent is obtained in writing from the relevant authority.
2 It is desirable to erect the cable above or below a nest such as this.

General Health and Safety Guidelines

There is little doubt that an approach that minimizes or better still, eliminates possible accidents or incidents is highly advantageous.

Responsibility of Management: It is the responsibility of management to ensure that all team members and supervisors are trained and familiar with applicable safe working practices, and that they take immediate and decisive action when safe and approved work methods are not followed.

Responsibility of Supervisors: It is the responsibility of the supervisors to ensure that each member of his team wears the required PPE and to ensure that the work area is protected by the use of the necessary signs, cones, flashing lights, traffic control personnel, etc. Personal protective equipment (PPE) refers to protective clothing, hard hats, safety glasses, or other garments or equipment designed to protect the wearer's body from injury. On top of this, practice safe and approved work methods, as generally outlined in this Manual.

Each Party must at all times comply with health and the safety legislation, regulations and guidelines, which must include, but is not limited to:

1. A competent person shall, before the commencement of any construction work, perform a risk assessment which shall be written into the health and safety plan to mitigate risks and shall include:
   a. Activity-specific hazard and risk identification.
   b. Assess and evaluate each identified hazard and risk and rank them i.e. high, medium or low.
   c. The best way to protect people is to eliminate the hazard or risk and second best, minimize it.
2. All areas used by the public shall be maintained free from debris or equipment that may constitute slipping, tripping, or any other hazard.
3. Adhere to all the health and safety management plan procedures.
4. Develop and obtain approval for a Traffic Management Plan (TMP).
5. Report and record all Work Site accidents, incidents and property damaged.
6. Establishing safe air space requirements prior to the use of lifting and construction equipment.
7. All personnel shall be required to wear the following personal protective equipment (PPE):
   - Protective overall (at all times).
   - As a general rule, steel-toed safety boots should be worn at all times.
   - Hard hat (when performing work that requires the use thereof).
   - Safety glasses (when performing work that requires the use thereof).
   - Work gloves help prevent cuts and bruises from sharp or rough edges on pipe/ducts and other objects.
   - Wear high-visibility vests (at all times).
8. The contractor shall ensure that all necessary guards, protective structures and warning signs are used to protect both workers and third parties. All necessary barriers and fences shall be erected to guide pedestrians and traffic around the work area.
9. A first aid box will be provided and allocated to a trained, certified first aider. Every injury occurring on site must be treated and recorded.
   - Should an injury require professional medical treatment, the supervisor in charge must complete an accident report.
   - Ensure that the first aid kit is available and accessible, correctly stocked and a register exists to account for used/missing items.
10. All employees, management personnel and visitors shall undergo induction training carried out by the Site Manager or a designated deputy before going onto site for the first time. Inductions records shall be kept on site for the duration of the project.

Typical Table of Contents for a Site Safety File

Site safety starts here

1. Notification of Construction Work
2. Letter of Good Standing
3. Organogram
4. Health & Safety Policy
5. Health & Safety Plan
6. Environmental Policy
7. Environmental Plan
8. Waste Management Plan
9. Fall Protection Plan
10. Emergency Plan
11. Emergency Contact Numbers
12. List of Sub Contractors
13. 37.2 & 5.3.b Legal Agreement
14. Appointments
15. Certificates of Competency
16. Risk Assessments
17. Induction Records
18. Toolbox Talks
19. Inspection Registers
20. Visitors Register
21. Complaints Register
22. Site Diary
23. Weekly Statistics
24. Safety Minutes
25. Audit Template
26. Vehicle List
27. Incident Records
28. Client's SHE Specification
29. Public Liability Insurance

Traffic Management Plan (examples)

1 No work should commence on a public roadway without first obtaining a wayleave from the road authority concerned.

2 It is the responsibility of the supervisor/s to ensure that each member of his crew wears the required PPE and to ensure that the work area is protected by the use of the various signs, cones, flashing lights, traffic control personnel, etc.

3 Traffic movement shall be inhibited as little as possible. Should this be unavoidable, alternative access to routes must be made available.

4 Work carried out on busy roads, should be restricted to outside the following periods; from 06:30 to 09:00 and 15:30 to 18:00, to ensure the free flow of traffic during peak hours.

5 Roads shall be kept free of debris or equipment.

6 Excavated material unsuitable for re-use shall be removed from site as soon as possible.

7 Where cyclists and/or pedestrians are likely to be present, their need for safe and convenient passage must be considered and sufficient, safe crossings shall be planned for.

8 Create 'no go' zones around hazardous areas and implement safe work distances.

9 Choose signs with messages clearly indicating the actions drivers or pedestrians are required to take.

10 Where necessary, traffic control persons shall be used to provide positive guidance to motorists.

11 Remember that the visibility of hazards/workers can be greatly diminished in darkness and/or poor weather conditions.

Selecting Signs

1 Choose signs that are appropriate; signs that accurately describe the work situation.

2 Start with general sign messages at the beginning of the work zone. Then use signs with more specific messages, stating what action should be taken, closer to the actual work area.

3 The overall effect of signs should be to make drivers aware of what they are approaching and what action(s) will be required of them.

4 Drive through checks should be made every so often, both at night and day, to ensure that signs are properly located to allow adequate driver response time.

5   Use only signs that appear in the local Road Traffic Signs Manual.

6   Signs must be kept clean and well maintained if they are to be effective.

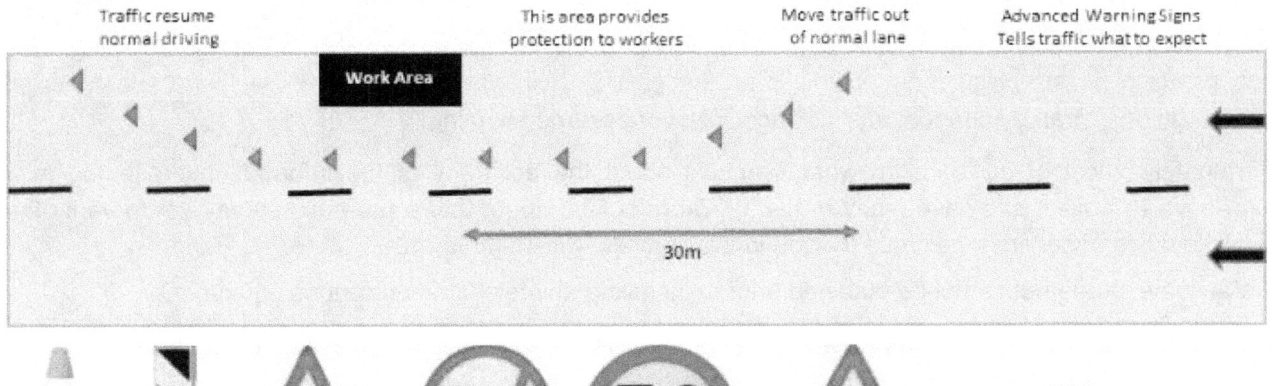

     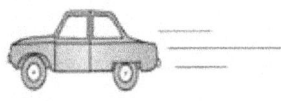

Flagging PPE and Communication

1   – A high-visibility reflective vest
    – A white hard hat
    – Steel-toed safety boots
    – Full length pants or coveralls - no shorts
    – During rainy weather, highly visible rainwear

2   When two flaggers are used, they can communicate verbally or visually if they are close enough to each other.

It is desirable to appoint one of the flaggers as the coordinator.

Where the end of a one-lane section is not visible from the other end, the flaggers must communicate via two-way radio.

The safety of workers and the travelling public, while passing through the construction area, depends on the efficient actions of flag persons.

3   A Warning Flag Signal may also be used to warn a road user to proceed slowly, and be alert of a hazard in or adjacent to the roadway ahead.

4   A good, active flag person can be as effective as any other means of drawing attention to a hazard in the roadway.

600mm x 600mm red flag
with 1m wooden pole

Wayleaves

A Wayleave is permission to use someone else's property to deploy infrastructure. In other countries this is also known as a right-of-way or an easement.

Wayleaves will indicate the positions of all other services. So, on "paper", we should not "hit" anything. This can prove to be as misguidedly optimistic as that of US Civil War general John Sedgwick's last words. "They couldn't hit an elephant at this distance." "Never assume anything!"

Fortunately, most of us are somewhat skeptical about the accuracy of the information presented in a Wayleave (a project can have many of them). There is little doubt that a pre-build survey set to verify the exact location of services indicated (or not) in Wayleave's, will do no harm.

A Wayleave agreement must be obtained prior to installing any telecommunications equipment.

Companies doing work in a Road Reserve shall at all times keep a copy of the Wayleave on site.

Companies must familiarise themselves with the Standards and special conditions as set out in this permit.

Companies are held responsible and accountable for the quality of work they deliver, as well as any objectionable actions by their workforce.

Pre-Build Procedures

A fibre installation project is a major undertaking. Responsibility for the oversight of everything from detailed implementation plans to community relations, ensuring sufficient materials are ordered in a timely manner to safety and environmental concerns, means an intense amount of pre-work and ongoing coordination for the life of a project.

1 As an unfaltering believer in a pre-build survey: The verification of details contained in pre-build drawings, will ensure that potential problem areas are uncovered before the contemplated work kicks-off and potentially save one a lot of trouble later.

2 Using the information available on the pre-build drawings, walk the pegged out route by foot, to determine the following:
   - Verify the soil classification/s by digging a pilot hole every 1km (hard, intermediate, soft)
   - If the soils or soil properties are not what were expected as noted in the contract, the client must immediately be consulted
   - Check and verify above and below ground utility locations
   - Note changes in gradient and/or direction
   - Identify all obvious landmarks where the route changes direction (take photos)
   - Take photos of all obstacles along the route
   - Verify HH / MH positions
   - Note road / rail crossings
   - Record crossings with other services
   - Record the presence of structures near the trench
   - Double-check the recorded details on the return journey
   - The route is typically marked using lime

Pre-Build Survey Equipment and Tools

1 Digital camera and spare batteries.
2 GPS with tracking function and spare batteries.
3 DCP tester.
4 Tape measure.
5 Measuring wheel.
6 Clip board, note book and stationary.
7 Route drawing/s from the client.
8 Reflector jacket.
9 Personal Identification.

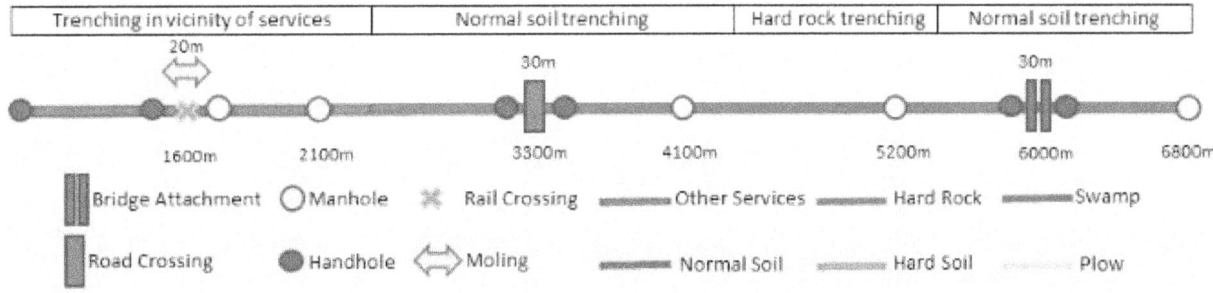

Armed with this in information, a meticulous build plan (productivity for day, week, month, etc.) can be developed and the contractor can ensure that the necessary equipment and recourses are in place.

Procedures and processes can now also be developed and put into place to ensure that the contemplated civil build is done to standards and executed within budget and timelines.

Contractor's Camp Establishment / Holding Area

1. On large projects, a contractor may need to provide for a construction site camp / safe holding area, which includes office accommodation.
2. The site location must be at a convenient point and as close as possible to the work site.
3. The contractor must provide for a safe holding area to store all material obtained from the client.
4. The contractor will bear all interrelated costs associated with securing the property and the camp establishment.
5. Must be fittingly fenced 1.8m high.
6. Must have secure lockable gates 1.8m high.
7. Must have a suitable office in compliance with OHASA and local authority requirements.
8. Must provide for sanitary facilities in compliance with OHASA and local authority requirements
9. Must contain telephone and Internet services (mobile and/or fixed).
10. The camp must be sufficient in size to accommodate all material and equipment required for the project.
11. On completion of the project, the contractor shall reinstate the camp establishment to its original state or better.

Trenching

It's immensely important for trenches to be excavated to such a depth that the crown of the duct has at least 800mm of backfill cover, in all soil conditions, except for where hard rock conditions are encountered.

Where it is not possible to obtain the specified minimum trench depth, the client must be consulted.

The trench depth in hard rock conditions can be relaxed (i.e. apply for a concession) to a minimum depth of 300mm backfill cover over the uppermost duct. But this being said, it now requires protection in the form of a concrete slab (either pre-cast, or cast in situ) placed on top of the padding material before backfilling. This concrete slab shall have a strength of 20 Mpa reinforced with high tensile wires and measure; 75mm thick by 300mm wide, and 900mm in length.

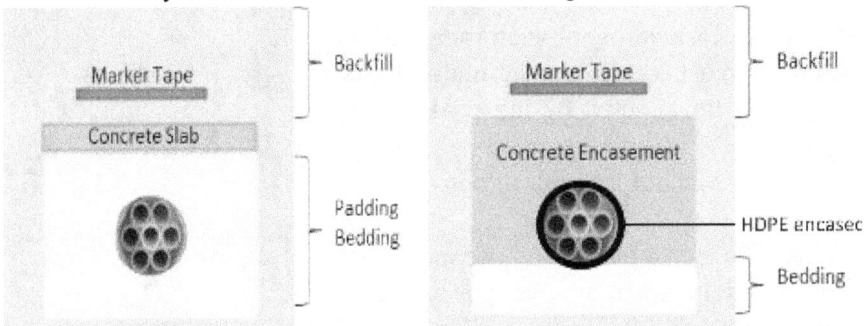

Concrete encasing is not endorsed enthusiastically by everyone; some argue that it turns a previously flexible duct into a long unreinforced concrete beam of little strength, prone to fracture with ground movement. And, this in turn could potentially damage an encased duct. A view not shared by everyone.

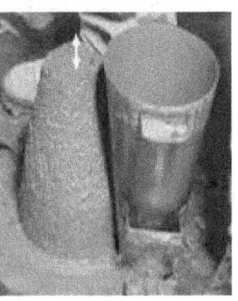

Before pouring concrete, a slump test must be performed (take photos of this procedure). How do we measure the ideal slump? A slump of 10 cm or less is typically deemed acceptable (must not shear-off or collapse) - or as per client spec. Concrete that is poured too wet will be weak, regardless of how it is cured

Trench Width

One other obvious consideration is the width of the trench, which of course, is dependent on the duct diameter.

Trenches that are too narrow will not allow for proper duct installation, whereas trenches that are overly wide are unnecessarily costly. On top of this, a too wide a trench will allow for too much duct snaking from the reel memory. Below are typical examples:

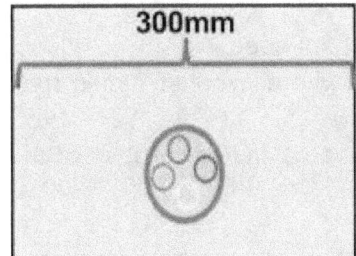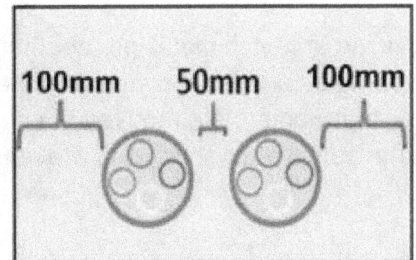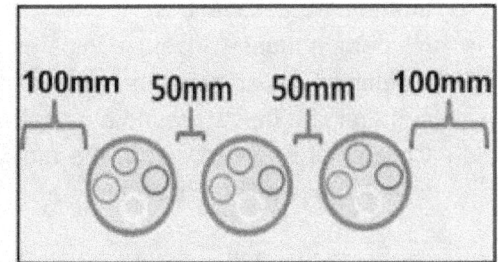

Areas where work is to be performed shall be cleared of all trees, shrubs, rubbish, and other objectionable material of any kind, which, if left in place, would interfere with the proper performance or completion of the contemplated work.

Pilot Holes

When the trench has been set out, Pilot Holes needs to be dug at 30 - 50m intervals, particularly at points where the new trench crosses existing services. The pilot holes should be at least 150mm deeper and wider than the proposed trench.

Pilot holes are one of the most effective methods utilized not only for the location of services, but also to determine the position of a trench, relative to other services.

Location of Services

1   In my view, investing a few shekels in a Ground penetration radar (GPR) is a ticket to fail-safe trenching. They are used to identify underground services and formations (readings can be affected by the presence of high voltage power cables).
2   No excavation work shall commence without having a copy of the approved wayleave/s on site.
3   It is the responsibility of the Contractor to locate all existing services.
4   A Utility Representative can be asked to point out the position of a service and sometimes even oversee the work.
5   Cable Locators can find the exact path and even estimate the depth of the utility service.
6   Hand excavating is necessary to uncover known services prior to commencing with mechanical excavation.
7   Any trenching done in the vicinity of existing services should be done very carefully to prevent accidental damage to a service.
8   The Contractor shall be liable for any damages to existing services.
9   Any damage to existing services must be reported immediately to the project manager.

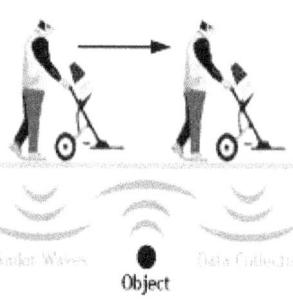

Trenches Deeper than 1.2 meters

1   Where the depth of a trench exceeds 1.2m and workers need to enter the trench, adequate measures must be taken by the contractor to provide support for this trench.
2   3 basic methods used for protecting workers against trench cave-ins are:
    -   sloping
    -   trench boxes
    -   shoring
    … and, a ladder provides a safe means of access
3   Barricades and signs are to be used at safe distances from edges to protect unattended excavations.
4   No person must work alone in an excavation or trench that is greater than 1.2m deep.
5   Machinery must never be placed in or near excavations and trenches where exhaust fumes may contaminate below ground atmospheres that workers are required to occupy.
6   Care shall be exercised in the moving or removal of shoring to prevent the caving or collapse of the trench faces being supported.

Notifications

1   Businesses / property owners shall be informed one week (7-days) in advance of any construction activities commencing in the vicinity of their properties.

2   These notices will announce upcoming work tasks and potential impacts, such as traffic, parking, and access changes, noise, utility interruptions, vibration, etc.

3   If a private driveway or footway constructed with non-standard materials is to be excavated, the owner of the property concerned must be informed in advance and in writing of the intended work.

Private Property

1   Where possible, excavations on private property shall not be left open outside normal working hours (08:00 to 17:00).

2   The Contractor shall be responsible for the protection of all trees, shrubs, fences, and other landscape items adjacent to or within the work area.

3   Occupants of the properties must be kept informed at all times of how their access will be affected.

4   When trenching through entrances to properties, access must be maintained by using steel plates or other temporary bridges of ample strength and, it must be well secured against movement.

5   Surfaces shall always be reinstated to the original state or better.

6   Where a Contractor must undertake tree and bush cutting and/or shrub clearing he must prior to undertaking such work, obtain approval in writing from the relevant authority and/or property owner.

7   The Contractor shall dispose of all cuttings and cleared material.

8   The Contractor shall be solely responsible and accountable to remedy any damages and/or claims, arising due to his activities.

9   All drainage systems must be cleared daily.

10 In residential areas the reinstatement of paving, grass or landscaping must be done to the property owner's satisfaction.

11 Remove all material and equipment not needed onsite, as soon as possible.

Suggested Trenching Practices (possibly not mentioned elsewhere)

1   All excavation work must be performed under the supervision of a responsible person who must be competent to exercise such supervision.

2   Proper excavation and preparation of the trench will inhibit unanticipated longitudinal and cross-sectional strains and stresses on the duct.

3   Trench walls shall be vertical for at least the height of the bedding and then as vertical as possible.

4   Exercise care when trimming trench floors to ensure that they are level.

5   If any street furniture (street names, traffic signs, bus shelters, etc.) have to be moved, arrangements must be made with the relevant authority for the removal, storage and re-erection.

6   Mark-out the proposed trench with lime.

Barricading

1   To begin with, provide adequate warning, guidance and protection for motorists, pedestrians, cyclists and workers from all foreseeable hazards.
2   In high traffic areas, erect fencing or place barriers measuring at least 1,2m in height -as close as possible to the excavation.
3   All barricading shall be well supported.
4   An additional precaution is to provide clearly visible boundary indicators at night or when visibility is poor.
5   Wear high-visibility vests (at all times), safety boots and hard hats when working at or near a public street or highway or when working at night.
6   When excavations are in progress next to the road, ensure proper and sufficient warning road signs placed at distances well before and after the work in progress. To warn approaching traffic, a flag person must be placed at least 40m before a "men at work" sign. Approved road cones must be placed at regular intervals along the whole route where trenching is in progress.
7   Backfill as soon as possible.

Trenching Near Kerbs, Guttering, Paving and Driveway Crossings

1   Where excavations pass beneath kerbs, guttering or driveways, etc., proper support shall be provided for these structures until tunnelling and backfilling is completed.
2   Where tunnelling is not an option, the existing concrete paving shall be neatly cut with an angle grinder to deliver smooth, uniform edges.
3   Where ducts are to be laid beneath existing paving blocks, the pavers shall be carefully removed to be reused. Paving blocks must be re-laid on a bedding of sand and reinstated to its original state or better.

Road Crossings

1   The contractor shall inform the relevant road authority 48-hours prior to the commencement of the work.
2   Directional drilling is the preferred method for crossing roads.
3   It is the responsibility of the Contractor to ensure that every law regarding traffic, safety, traffic signs and barricading is complied with.
4   The angle of the crossing should be as near a right angle to the road centreline as possible.
5   The edge of the trench must be cut using asphalt/concrete cutters to deliver smooth, uniform edges.

6   The minimum depth that any service may be placed under a road is 800 mm.
7   The duct/s shall extend a distance of at least 0.5m beyond the edge of the road.
8   All excavated material and equipment must be placed and demarcated in such a way to not inconvenience vehicles and pedestrians.
9   No person may off-load on a public road, any materials that are likely to cause damage to a road surface.

Stream and River Crossings

1   Before any serious thought is given to the installation of a high-density polyethylene (HDPE) duct at a river crossing, the network designer will consult with a geotechnical expert to conduct a comprehensive study.
2   This investigation will reveal the most efficient way to accomplish the crossing in question. Horizontal directional drilling (HDD) (discussed later), has become a popular river crossing option, and as explained later, rightfully so.
3   The duct must be sealed at both ends to prevent water or dirt ingress.

Bridge Crossings

1   This work is typically undertaken by subject matter experts in this field.
2   First and foremost, the use of existing ducts or service culverts within bridges must be fully explored.
3   The contractor shall inform the bridge owner 48-hours prior to the commencement of the work.
4   Ducts attached to the underside of bridges must not affect its load bearing capacity, reduce the clearance or cause other issues.
4   Not all bridge structures will have the exact same installation configuration and procedures may vary to accommodate your specific requirements.
5   The most common method to be used will involve the use of a hydraulically operated crane fitted with a safety basket be positioned adjacent to the bridge balustrade. From this position, workers wearing safety harnesses can be hoisted over the balustrade and lowered into a working position as required.
6   Bracket mounting positions can now marked out on the side or underside of the bridge as directed by the design drawings and instructions. Next, drill the holes, fit the concrete anchors and mounting brackets and firmly secure them.
7   The steel or ultra-high-density polyvinyl chloride (UPVC) base carrier duct can now be positioned and firmly secured.
8   The microducts can then be hauled through the newly mounted base carrier. Use a continuous length of duct (no joints permitted).
9   Where required and as stipulated in the design instructions, both the approach and departure ends may have to be encased in concrete where they traverse the bridge abutments and enter the ground. It is desirable of course, for the end-product to be both safe and visually appeasing.

Trenching Near Power Cables

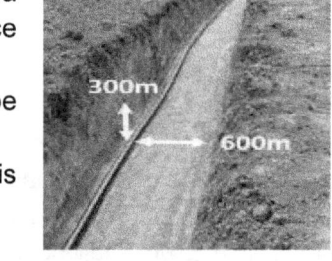

1   Where no physical barrier exists, no duct or cable shall be laid within a distance of 600mm measured horizontally, nor cross within a distance of 300mm measured vertically from any high voltage power cable.
2   Where this separation is compromised, the duct or cable must be separated by concrete slabs.
3   The standard protection slab is 900mm x 300mm x 75mm thick. This slab will be reinforced with 3.55mm high tensile wires.

Rock Blasting

1. Blasting for excavation shall not be performed without written permission obtained in advance, from the agency having jurisdiction.
2. Procedures and methods of blasting shall conform to all local laws and ordinances.
3. It is the responsibility of the Contractor to establish appropriate safety and health practices and determine the applicability of regulatory limitations prior to blasting.

Duct Deviation

1. HHs / MHs typically facilitate changes in direction.
2. However, ducts may deviate from a straight line provided that:
   - Individual lengths may be offset by no more than 35mm.
   - The offset is in the same direction, this to avoid creating S-bends.
   - The maximum overall deviation between MHs / HHs does not exceed 15°.
3. Spacers should be used when placing multiple ducts in a trench. They prevent ducts from twisting over and around each other. By keeping the ducts in straight alignment, cable jetting and/or pulling tensions will be reduced.

Steep Gradient Trenching

1. Where excavations needs to be done on up-hills or declines, a layer of sand bags should be placed at regular intervals (5m apart), to prevent the possible wash away of backfill material.
2. Always consider the risks of land stability or earth movement, when trenching on embankments (specialist technical investigation may be required).

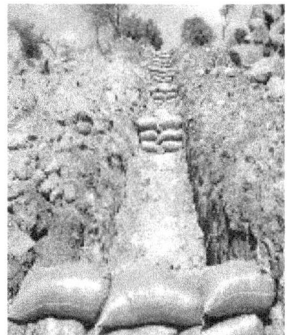

Tree Roots

1. Tree roots exposed while trenching must not be cut unless absolutely unavoidable.
2. Prior to undertaking tree/bush/shrub clearing/cutting as may be necessary for trenching, approval in writing from the relevant authority and/or property owner needs to be obtained.
3. Never cut roots over 25mm in diameter, unless advice has been sought from the local authority.
4. Ducts are to be sleeved in a HDPE pipe, if not 100mm away from existing tree roots.

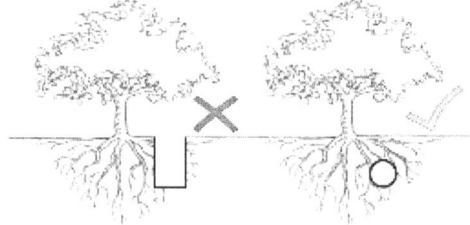

Surface Material

1. Surface material (paving slabs, soil, grass, etc.) must be kept apart by placing them on opposite sides of a trench, where they are least likely to interfere with traffic, pedestrians and drainage systems.
2. Mow-able lawn shall be cut in square blocks and put aside and be kept moist until reinstated.
3. Where trenches pass through gardens, the contractor shall seek direction from the owner.

 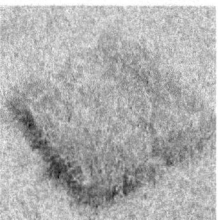

Duct Un-Coiling and Installation Process

1 Ducts shall not be un-coiled without the use of a Vertical or Horizontal De-Coiler. De-coilers will prevent twisting, bending or kinking from occurring during the installation process.

2 Duct un-coiling can be accomplished by pulling the conduit straight into a trench from a stationary rotating de-coiler or by laying the conduit into the trench from a forward moving de-coiler positioned on a trailer.

3 Once the duct coils are secured inside the de-coiler, only then can the containment straps on the duct coil be cut. Next, rotate the de-coiler slowly to unwind the duct out in one plane.

4 Generally, the ducts are placed in the trench, one length at a time and joined on the floor of the trench using couplers.

5 Ducts shall be laid in a straight line between MH/HHs. It is never ideal to have directional changes, but if unavoidable - keep the bending radius as big as possible and offset is in the same direction.

6 As the ducts are laid and jointed, install end caps on ducts at all MH/HHs to prevent water and dust from entering.

7 Care shall be taken to ensure that no dirt collects between the duct and coupler to deliver an airtight seal.

8 At MHs or HHs where the duct goes straight-through, allow for sufficient slack for the duct to be secured against MH or HH walls.

Duct Installation - Moving Trailer Method

1 This method is most efficiently used where the path to be followed does not contain any obstructions that require the duct to be placed under. Move the trailer slowly along the trench route while unwinding and placing the duct in the trench. Take care not to over spin the reel.

Duct Installation - Pulling Method

1 The duct can also be pulled and placed by hand or by a mechanical pulling machine with the help of a Pulling Device that is fitted in-between duct and mechanical pulling machine. The two types of pulling attachment devices most commonly used are a Pulling Grip or Basket Grip.

Bedding and Padding

Bedding & Padding

Backfill

1 Bedding is the material constituting the even (rake if necessary) floor of an excavated trench onto which ducts are laid.

2 It is desirable to pass both bedding and padding through a sieve. The material used for bedding and padding must be of a granular, non-cohesive nature, graded between 0.6 mm and 13 mm, or as per client spec.

3 Care shall be taken to place padding material simultaneously on both sides of the duct to prevent lateral movement.

4 The compaction of padding shall be executed manually using a hand tamper. Duct buckling is much more probable when the padding material does not provide adequate side support.

Backfill and Warning Tape

1 After padding tampering, backfilling of the trench can be done.
2 Material excavated from trenches may be used as backfill, provided that it contains stones no greater than 150mm in diameter, trash, or organic matter that could potentially damage ducts.
3 Backfill material is to be installed in layers not exceeding 300 mm, with each layer compacted before the next is added.
4 After compacting the first layer of backfill, the warning/marking tape is placed. Take photos of this procedure as proof of existence. Conceivably, the warning tape will be encountered before damaging any ducts or cable.

Backfilling (Concrete)

1 Check the consistency of the concrete (slump test).
2 Tamper the concrete using special care not to damage the ducts.
3 Check for cavities.
4 Allow for the concrete to cure (get hard).
5 As mentioned earlier, concrete backfill turns a once flexible duct into a long unreinforced concrete beam of negligible strength, very likely to fracture with ground movement. This in turn could potentially damage the duct. I ought to point out though, that this view is not shared by everyone.

Trench Compacting

1 Once a poorly compacted layer is in place, it is difficult if not impossible to achieve good compaction in the layers above. This is a key point; a consequence of poor backfill compaction is high % air voids. This potentially makes the duct vulnerable to veld fires.
2 Manual compaction is performed until the ducts are covered by both a 150mm layer of padding and 300mm of backfill, at which point a vibratory plate compactor can be used.
3 The compaction of the final backfill layer shall be by means of a compaction machine and shall be compacted to a density higher than or at least equal to that of the virgin soil parallel to the trench.
4 After completion of the backfill, a DCP test must be done. This test must be documented.

Optimum Moisture Content (OMC)

1 Moisture conditioning of the backfill material shall be carried out by the contractor. If not specified by the client, the following 2 Field Tests for Optimum Moisture Content (OMC) can be used:
2 A handful of backfill tightly squeezed in the hand, shall be wet enough so that it binds together with no more than slight crumbling when the hand is opened.

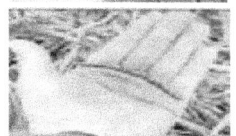

Dynamic Cone Penetrometer (DCP) Testing

1. All excavations are subject to compaction tests - which must be documented.
2. 8 tests per km, is recommended.
3. Uses a free-falling 8-kg hammer which strikes a cone, causing the cone to penetrate the soil, and then measuring the penetration per blow, called the penetration rate (PR), in mm/blow.
4. Always keep the DCP vertical and watch where you place your fingers.
5. The trench density must be better or at least equal to that of the virgin soil parallel to the trench.
6. A 25mm PR is typically deemed acceptable.

Direct-Buried Installations

On paper, direct-bury is fairly simple and very cost-effective - you dig a trench, install the cable and re-instate. However, once in the ground, it is perhaps more susceptible to damage from unrelated digging activities than cable in a duct, and much more importantly, it is difficult to access and repair. In Africa, these issues are greeted with a philosophical shrug.

1. On all direct-buried installations, Armoured Cable must be deployed - which provides for both crush and rodent protection.
2. In general, the most desirable and economical method of cable placement in open or rural areas is ploughing- where there likely to be fewer obstacles to impede the progress of the ploughing equipment. In urban or sub-urban areas where there can be many obstacles such as underground utilities, sidewalks, road crossings etc., trenching by hand has its advantages.
3. Should only an armoured cable be placed, a trench no wider than 375mm will do.
4. The trench shall be1.2m deep or as stipulated in the wayleave.
5. Grade off abrupt changes in terrain ahead of the plough.
6. Excavate at least 5m at the starting point, to allow for the plough to immediately lay the cable at the correct depth.
7. Always start the plough tractor's movement slowly and increase speed gradually only after the cable slack is taken up from the cable delivery system.
8. Ploughing operations must be observed continuously for: obstructions, proper feeding of the cable, specified depth, following of the marked route, and the safety of the crew.
9. Stoppages in the plough-in process should only be for the loading of cable or marker tape and when the terrain demands for this.
10. Warning Tape must be placed approximately 300mm below the ground surface, directly above the cable.
11. It is critically important for MHs and HHs to be built only after the plough in of the cable. Pre-fabricated HHs are often utilised.
12. Each section, after ploughed-in, must be checked with an OTDR for possible damage.
13. After the plough in process is completed, the trench must be levelled by one of the following methods:
    - Back blading with the plough-in machine.

– Using a TLB to level the disturbed areas.

14 The National Electrical Code (NEC) recommends that non-current carrying armour shields and metallic strength members be bonded and grounded.

Earthing, Bonding and Surge Protection

1 The armouring of optical fibre cables shall be lugged and bonded to an earth bar using a soft multi-stranded 6mm² green / yellow insulated bonding cables. Bonding cables shall be kept as short as practically possible and must contain no sharp bends.

2 Proper grounding and bonding is required for the safe and effective dissipation of unwanted electrical current, and it promotes personal and site safety. Typically, optical fibre cables do not carry electrical power, but the metallic components of a conductive cable are capable of transmitting current.

3 When the armoured cable is correctly bonded and grounded, it minimizes the risk of unwanted electrical current that could potentially harm personnel, property or equipment.

4 Required...By electrical codes and equipment manufactures | To protect personnel and equipment

The difference between the terms bonding and grounding:

1 **Bonding** is the multiple connections to metallic parts (at every joint), required to form a continuous electrical path.

2 **Grounding** is the act of connecting that path to an earth.

The Procedure

1 Use a cable knife to score the outer sheath of the armoured cable approximately 25 mm (1 inch) long. Take care not to damage the inner sheath.

2 Slide the base plate under the armour.

3 Place the top plate over the base plate and tighten down the lock nut.

4 Cover the grounding clamp and split portion of the sheath with vinyl tape and connect to the ground system or bond thru.

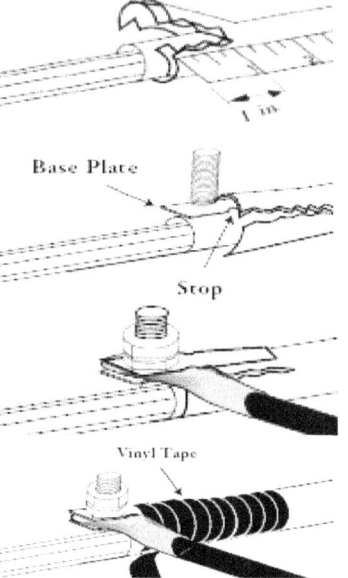

Markers

1 In 'non-built-up areas' underground routes must be marked with identifiable markers.

2 GPS coordinates of all MHs /HHs must be taken and documented to form part of the as-built documents.

3 These markers shall have a length of not less than 1.8m and a diameter of not less than 100mm.

4 Markers shall be planted 600 mm deep, opposite a MH or HH and be well compacted.

5 Passive markers can also be buried at key points during construction or used to mark existing facilities.

Manholes and Handholes

1. MHs and HHs shall be positioned as far away as possible from road junctions.
2. MHs and HHs must to be built according to prescribed dimensions and specifications.
3. Before any concrete is placed the Contractor shall examine the shuttering for firmness and correctness of position and remove all dirt and other foreign matter.
4. Hand mixing of concrete is permitted in exceptional cases but only with written permission of the client.
5. Concrete mix shall be such that a strength of 20MPa (2.5:2.5:1) is obtained 28 days after pouring.
6. Duct entry points into HHs / MHs must be drilled, without cracking or damaging the surrounding structure.
7. Ducts shall enter and exit HHs / MHs in line with the direction of the route, for them to be coupled thru without any obvious effort, as a continuous duct.
9. HH / MH external labeling should be done on the coping and NOT the lid, as lids can get damaged and be replaced. GPS coordinates must be recorded.
10. On completion of a HH / MH, the Contractor shall re-instate the area around the HH / MH to its original state or better.
11. HH / MH covers shall be finished flush with the surface area.

Where bricks are used to construct HHs the following must apply:

- Written permission obtained from the client.
- Use only baked quality clay bricks sourced from an approved manufacturer.
- Re-enforce every third layer of bricks.
- The wall thickness must be double brick or better.
- The outer wall must be waterproofed.
- Use an approved footway-type or roadway-type frame and cover.

HH / MH Inspection

1. Splice closures are secured.
2. Cable slack neatly stored and secured with no compromise to the bending radius.
3. Left tidy and clean.
4. Ducts sealed in-between wall.
5. End Caps fitted to empty ducts.
6. Used ducts sealed in-between cable and duct.
7. Cables and ducts are clearly labelled.
8. HH/MH clearly marked on the coping.
9. Locating disk in position.

HH Installation

1. All duct entries and exits at the HHs must be a watertight seal.
2. All ducts in HHs shall be coupled through.
3. Incoming ducts must have a watertight seal.
4. Ducts must be sealed with a watertight coupling that is cast or inserted into the wall of the HH
5. HH covers must be watertight or must have at least a double seal.
6. HH covers should be 150mm above natural ground level with the fill shaped back to natural ground
7. level in a 2m radius around the HH cover.
8. On paved sidewalks or verges, next to roads or streets, a cast in-situ concrete or asphalt backfilling shall be sloped to not impede pedestrian traffic. In these instances the HH installation shall be such that it is not more than 50mm higher than the surrounding paved sidewalk.
9. The inside surface of the HH shall be sealed using an approved bituminous product.

Safe HH / MH Working Procedures

1. Obtain a HH / MH entry ref for the work to be undertaken.
2. At least 2-persons must be present before entering a MH.
3. Ensure that the vehicles are parked in such a manner that they do not create an obstruction or hazard to traffic and/or pedestrians.
4. Use barricades and cones that are clearly visible around the HH / MH.
5. Pour water around the lid of the HH / MH, to prevent creating a spark when opening it.
6. Lift HH / MH covers using your legs and place the cover at least 2m away from the opening.
7. Before entering the confined space, test at 3-levels (bottom, middle, and top) for:
   - Oxygen content
   - Flammable or explosive gases
   - Hydrogen sulphide
   - Each level must be tested for a minimum of 60 seconds.
   - Use only an approved calibrated gas detector.
8. Ensure that the gas detector is in operation the entire time spent in the MH.
9. An aluminium ladder in good operating condition must be used to enter the MH.
10. Raise or lower tools and/or equipment into a MH using a rope or bucket.
11. Never place equipment or tools near the edge of a HH / MH.
12. Constant ventilation is required when performing work within a MH.
13. Water in a MH (not containing object-able content) shall be pumped-out into the storm water system. If not possible, or onto area, a suitable distance away, with positive drainage.
14. Never fusion splice in a HH or MH.

Trenchless Methods

Horizontal Directional Drilling (HDD)

1. In very simple terms, HDD is a drilling process where a drill head is steered underground. This method is adopted by all companies but habitually, sub-contracted to specialists.
2. HDD is the preferred method to cross roads, highways, railway lines, rivers and all other services that may prove to be too dangerous or costly to cross using conventional methods like trenching and/or ploughing.
3. The depth of any hole drilled for the installation of a new service, must be at least 800mm below surface of the road, or as per client spec.
4. In what follows, a brief overview on the 3 installation stages:
   - I. Pilot drilling
   - II. Reaming the initial pilot hole
   - III. Pulling back the duct in the reamed hole
5. The course of the drill is monitored and can be controlled as rods progress following an upward sloping path, before emerging at an intended point.
6. The drill head is then removed and replaced with a back reamer, ± 20% larger than the duct or cable to be pulled into the hole.

   The duct is attached to a swivel connection on the back reamer. The drill-rods and reamer are rotated and pulled through the hole, enlarging it and pulling-in the duct at the same time.

   The whole operation is carried out with pressurized drilling mud, which both carries away the spoil and supports the hole.
7. Rigs capable of drilling up to 300 metres in one drill are available and various sizes of ducts can be installed with this equipment.
8. The covering must not be less than three times the final diameter of the drilling hole and at a minimum of 1.5m.
9. At river crossings the distance between the bottom of the water and the drilling hole should be 10-times the diameter of the pipe and not less than 3m.
10. Soil removal during the drilling process is the responsibility of the contractor.
11. If the accuracy of the drilling is not specified in the wayleave, the area in which the drilling may wander should not exceed a 40mm diameter around the predetermined axis.
12. As always, the location and depth of underground services must be pre-determined before drilling can commence - as sudden deviations are not possible to bypass obstacles.

Reinstatements

1. Reinstatement work must be done by The Roads & Storm Water Department unless a wayleave indicates otherwise.
2. Should the wayleave holder do the permanent reinstatement, a 12-month guarantee period commences from the date of completion.
   OR
   Should the wayleave holder do a temporary reinstatement, a 2-week maintenance period commences from the date of completion.
3. Grassed areas shall be reinstated using the original turf, replacement turf or an equivalent seed.
4. Any constructed footway must be reinstated with the same surfacing materials that existed originally (e.g. concrete blocks, slabs, etc.). Material may be reused if undamaged, or else replaced with similar material.
5. Remedial work will be required if any of the following defects exists: depressions (resulting in standing water), humps (crowning), edge depression or cracking.
6. The performance of any trench permanently reinstated by the authorized person will be monitored for twelve (12) months, during which period the authorized person will be held responsible for any

remedial work that may be required.

7   Any excavation left unattended for a period of more than 5-days can be made safe by the road authority and charged to the contractor.

Road Reinstatement

1   The permanent reinstatement of the surfacing typically consists of 100 mm hot-mix asphalt. The lower 70 mm must be compacted and rolled asphaltic black base (26.5 mm nominal stone size, continuously graded) and the top 30 mm (4.75 mm nominal stone size, continuously graded). Cold-mix may be used only for temporary reinstatement.

2   The reinstated surfacing must be at least 100 mm wider than the trench on both sides to accommodate any edge breaks.

3   On completion of the work concerned the authorized person must fill in a completion notice and return it to the road authority within twenty-four (24) hours.

4   The road authority will then arrange a site meeting to do an inspection and to issue a certificate of completion if all requirements have been met.

5   A twelve (12) month guarantee period for permanent reinstatement by the wayleave holder, or the fourteen (14) day maintenance period for temporary reinstatement by the authorized person, commences on the day after the issue of the certificate of completion.

6   Rubbish, and other objectionable material of any kind, must be legally disposed of, absolving the client from any liability connected therewith.

Cable Pre-installation Procedures

The following items are key considerations:

Pre-test with an Optical Time-Domain Reflectometer (OTDR)

1   All optical fibre cables must be tested while on the drum, prior to deployment. A Bullet or Divot Bare Fibre Adapter is to be used to connect fibres to the OTDR.

2   Testing shall be done on all fibres in one direction at 1550nm, using a pulse width of 30ns. Traces must be stored and an electronic copy submitted to the project manager.

Pre-installation Cable Inspection

1   Always wear protective gloves and safety boots (with steel toe caps) when handling drums.

2   Check that the cable specified has been procured.

3   Inspect the cable drum for signs of excessive weathering and/or damage.

4   Drums must be transported or stored with their battens intact.

5   Do not accept delivery of an optical fibre cable should the drum is damaged.

6   The plastic foil wrap must remain in place until cable placement.

7   When removing the plastic foil wrap on the cable, do not use sharp tools that could potentially damage the cable jacket.

8   Ensure that the cable drum bolts are all tightened.

9   Verify that nails, bolts or screws on the inside surface of drum flanges are counter-sunk to avoid damage to the cable during placement.

10   Place the cable drum in line with the intended direction of deployment, in order to prevent the cable from rubbing against the reel flange.

11 Cable ends must always be sealed – using pre-formed or heat shrinkable end caps. Using tape for sealing cable ends is considered unsuitable.

12 Should the drum be rolled for some reason, always do so following the direction of the arrow.

13 Drums on site, must be chocked to prevent them from moving or rolling.

Cable pay-off - over the top or from the bottom? "Oh, oh, I know".

- Jetting / blowing cable
- Hauling cable
- Figure-eighting cable
- Erecting cable

As for me, over the top makes much more sense. However, advocates of from the bottom, when jetting / blowing cable do exist. This simply reinforces our conviction that ultimately, faced with this choice, a Contractor must determine which option is favoured by the Client.

Introduction to Air-Assisted Installations

Few (if any) would dispute that the foundation of the value proposition associated with blown fibre systems, is the low cost fibre upgrade capability it offers i.e. adding fibre cables to unused micro-tubes.

Jetting and **blowing** are two probable air-assisted cable installation techniques. Both methods require pushing the cable with a tractor mechanism while blowing compressed air into a preinstalled duct along the cable being installed. Both rely on air flow to help "float" the cable inside the duct and a result, minimizing sidewall pressures and reducing friction between the cable and the duct.

When jetting or blowing, you're combining a pulling force (compressed air) with a pushing force (hydraulically or pneumatically driven tracks) during the installation, providing an efficient, stress-free deployment in far greater increments than possible when hauling cable.

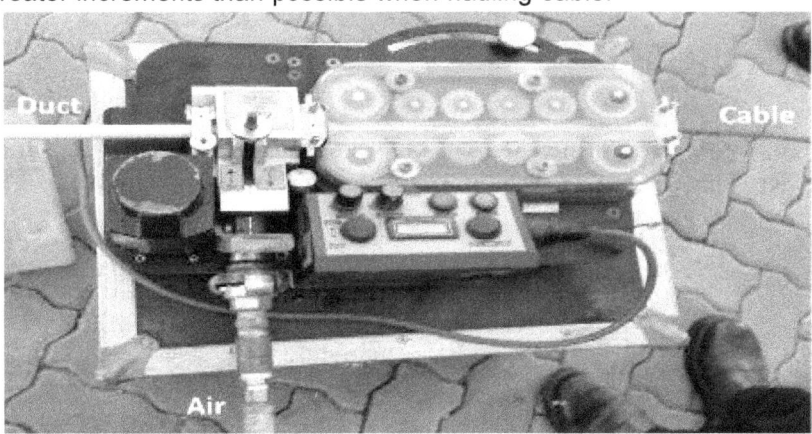

Jetting and blowing differ in how pulling force is applied to the cable. Jetting uses a missile at the front end of the cable. A differential pressure across the missile head creates a pulling force on the cable. Blowing does not use a missile - instead, the pulling force on the cable is due to fluid drag of air rushing along the cable - this pulling force is distributed along the cable length. A mechanical pushing device is common in both methods.

It is common wisdom that hydraulics can give you higher forces than pneumatics. Mindful of this, a blowing machine with a pneumatic drive is recommended for cables up to 15mm in diameter and a blowing machine with a hydraulic drive is recommended for cables with a diameter in the range from 14 - 32mm.

The key parameters when installing fibre cable in ducts are blowing distance and time. 1000 to 2000m is a typical blowing length. Typical installation speeds are 40-50 meters per minute under an air pressure of 10 bars. Most routes can be blown with a 10 bar supply.

Smaller diameter and lower weight cables make possible larger blowing distances. The maximum cable push force will decrease as the duct inside diameter increases, reducing the achievable blowing distances.

Survey the complete route prior to blowing in order to determine:
1   The location of HHs / MHs.
2   Determine distances between HH and MHs.
3   Determine where coupling of ducts are necessary.
4   Accessibility of terrain for the use of the intermediate (mid-point) fibre blowing equipment.
5   Identify specified ducts to be utilised.
6   Be sure to consider the accessibility of HHs / MHs to the splicing vehicles.

Jetting / Blowing the right cable

1 Choosing the right cable is extremely important for any installation. Strictly speaking, the purpose of the cable is to protect the fibres during installation and the service lifetime. Cables share some but not all characteristics and you need to ensure you install the cable type appropriate for the application.

2 Micro cables are designed with high-density polyethylene (HDPE) outer sheaths to minimize friction with the inner surface of micro ducts. They are also designed with the necessary stiffness properties to resist buckling forces and to negotiate changes in direction.

3 The cable on the drum must be covered until just prior to installation to protect the jacket from exposure to the sun.

Suggested Air-Assisted Installation Practices

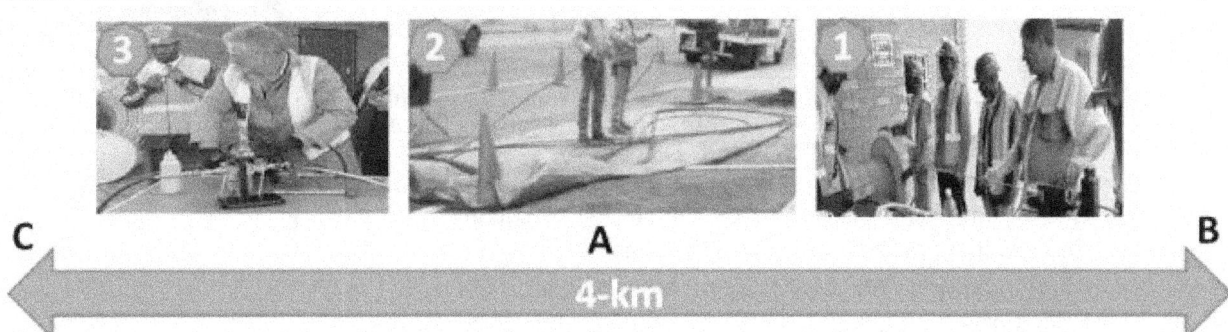

C   A   B

4-km

Cable drums are typically supplied containing 4-km of cable. Mindful of this, drums are placed midway of a 4-km length. There's plenty of evidence suggesting that the most efficient cable installation would involve two 2-km jetting shots (A to B followed by A to C). After jetting A to B, a 'figure-8' will be created at the midway point, to gain access to the end point of the cable previously on the drum. Next, the figure-8 end will be jetted in the other direction. This figure-8 puts a half twist in on one side of the 8 and takes it out on the other, preventing twists.

With a little bit of good fortune, we ought to be able to install 4-km of cable in one placement of the cable drum and blowing equipment... consisting of; 2-km of cable jetting - 2-km of "figure-eighting" - concluded Adding more air to a microduct further down, can also increase the achievable distance.with 2-km of cable jetting. If so, great. But wait, what if at some point the cable movement slows down during the blowing process - then instead of taking a 2-km shot, the jetting shot could for example, be reduced to only 1-km. This of course, would call for a second 'figure-8' to be made at the 1-km mark.

Blowing distance is directly related to the weight of cable, the pressure used and, friction from the inside of the microduct. Cable can be successfully blown further than 2-km when blowing heads are used in tandem.

   Maximum achievable distance   

Additional air can also be inserted further down, to achieve more distance.

Duct Integrity Testing (DIT)

All team members MUST wear eye protection during DIT testing as small particles can be blown out of ducts through the holes in the DIT catcher.

1 **Air Test**
   - Fit a DIT catcher on the far side and equip personnel with two-way radios.
   - Allow the air to flow through duct for at least one minute, to remove all loose particles and/or moisture.
   - Now proceed with the sponge test.

2 **Foam Sponge Test** - cleaning the duct:
   - A sponge is typically 100mm in length and 2 x the duct ID.
   - Wet the sponge slightly with blowing lubricant.
   - Place the tight-fitting foam sponge inside the microduct.
   - At a pressure of 10bars, blow the sponge through the microduct.
   - If excess water or dirt exits the microduct, repeat the process.
   - Have a DIT catcher in place at the far end to catch the sponge when it emerges.

Sponge

3 **Mandrill Test** - check for bends, kinks or blockages:
   - The sponge test MUST precede this test - a mandrill can damage a dirty duct.
   - Use a 40mm long mandrill made from Nylon or Teflon.
   - The OD must be no more than 85% of the microduct ID
   - At the receiving end, a DIT catcher must be used.
   - Note that a flying shuttle can cause injury and/or damage!
   - Always inspect the condition of the emerged mandrill, visible grooves is an indication of duct indents.

4 **Pressure Test** - check for coupler leaks or microduct punctures:
   - Fit a high-pressure end-cap to the duct under test on the far-side.
   - Gradually build the pressure up to 10 bars.
   - Test all coupling used for this test for leaks, using soap, water and a sponge.
   - Connect the air feed and leave this open until the pressure in duct stabilizes at 10 bars.
   - Close the air valve on the test assembly and monitor the pressure gauge for 5min.
   - Losing 1 bar in 5min is acceptable – any leak greater than that, must be found and fixed.

If the duct fails DIT tests, consult with the relevant authority on whether to use an alternative duct or to repair of the designated duct.

Micro Duct Fill Ratio

1 For optimum jetting / blowing performance, ensure that your cable-to-duct diameter fill ratio does not exceed 76%. This is determined and provided by the microduct supplier. It is wise to always verify this by checking the provided spec sheet.

Cable Blowing Lubricants

1 Adding a small amount of lubricant to a tight-fitting foam carrier (sponge) before blowing it through the duct, will provide an open and lubricated pathway. It is argued to give better results compared to coating both the duct and cable with lubricant. Odd, huh? Let me try to explain. Lubricant on a cable jacket apparently makes it too "slippery" for optimal laminar-flow jetting.

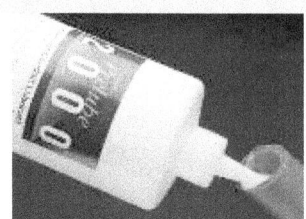

2 Let's not forget that the air "pushes" on the jacket through friction. A too low friction jacket will not easily be pushed by air. Normally, as little as 0.25 litre of lubricant is required to coat 2km of cable.

Suggested Air-Assisted Installation Practices

1 It is desirable to blow downhill wherever possible.
2 DIT results must be available prior to blowing the cable.
3 Always measure the cable diameter before jetting, to select the correct spacers. The cable must lie snug in the V-groove between the drive rollers.
4 To ensure that the cable is not prone to slipping, push and pull the cable through the machine by hand a few times to ensure that the cable is seated properly in the drive rollers.
5 The cable must always enter the jetting machine under no tension and in a straight line.
6 The cable must be clean as it enters the blowing equipment to allow for effective gripping of the driving wheels. Contamination of the cable will cause unwanted friction and result in a reduced blowing distance.

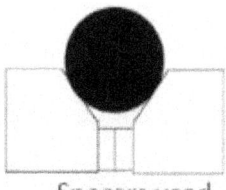

Spacers used

7 Monitor the speed and torque of the blowing machine as the cable is blown to ensure that the cable jacket is not damaged by the driving wheels.
8 One person must control the cable reel and be prepared for emergency stops.
9 Always note the meter reading on the cable and zero the blowing machine distance meter, prior to blowing.
10 Always establish two-way communication during the fibre deployment process at crucial points along the cable route.
11 The blowing machine should have an automatic shut-off feature that controls the push force applied to the cable. With this feature, cable and duct damage can be prevented should the cable stop abruptly within the micro duct.
12 Ensure that when connecting the transportation duct to the blowing machine, it is not entangled with the new or existing cable/s.
13 The 'figure-8' must be laid on a ground sheet and care must be taken to ensure that the cable does not touch the ground as it is fed into the machine.
14 To protect the end of the cable and also to guide the cable along its desired route, a bullet must be used.
15 As mentioned, blowing lubricant reduces the frictional drag on both cable and in the duct thus increasing the achievable distance.
16 To ensure that the blowing equipment remains in good working condition - after a week of continuous blowing - machine maintenance must be performed.

Management of Cable Slack

1 When handling cable, always wear; a protective overall, gloves and safety boots.
2 Never exceed the recommended cable bend radius:
   10 x Cable OD - No Tension (installed)
   20 x Cable OD -Under Tension(being installed)

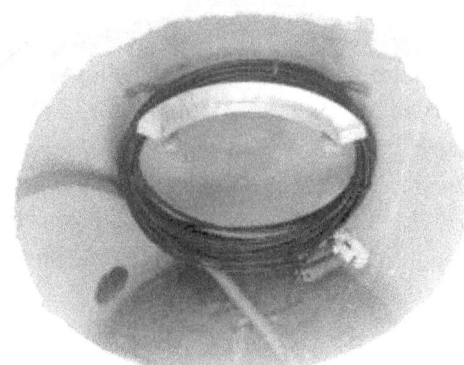

3. What is the minimum bend radius of your microducts?
   There is a simple formula: Radius min =10 x OD
   Therefore, if you have microduct has a OD of 10mm, your Radius min =100mm
4. A total length of 17m (15m for slack and 2m for splicing) of cable slack is traditionally required in a HH or MH housing a splice enclosure. Be sure to consider the accessibility of HHs / MHs to splicing vehicles.
5. Install slack brackets on HH / MH walls to secure cables.
6. The slack in the HH or MH needs to be tied together using a tie wrap or PVC tape at 1m intervals.
7. The fibre cable slack inside the HH / MH must be coiled in a 'clock-wise' direction with minimal back tension.
8. Ensure that the slack of one splice closure does not become inter-twined with the slack of other splice closures in the HH / MH.
9. Bundle cables together in groups of relevance.
10. Do not route cables in such a manner that they block ducts.
11. Used ducts must be sealed between cable and duct.
12. Cable slack typically adds 2% to the overall distance. Thus a 100-km link is likely to contain as much as ± 102 Km's of cable.

Figure-eighting

1. The preferred size of a figure-8 is about 5m in length, with each loop about 2.5m in diameter. To start with, set up two traffic cones 5m apart – then pay the cable off the top of the reel and loosely weave it around the cones in a figure-8 pattern - on top of a groundsheet. Large relaxed loops will help prevent the cable from becoming entangled.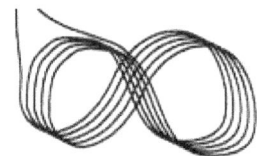
2. Continue to figure-8 the cable until the remainder of the reel is played off.
3. In order to pull from the figure-8 - it is necessary to flip it over, a task that requires a minimum of three people -one at the centre and one at each end. The required cable end will now be at the top of the figure-8.

Micro Duct Tools and Accessories
Tools
Ensure that installers are fully educated on the proper use and operation of any tool before starting a job.

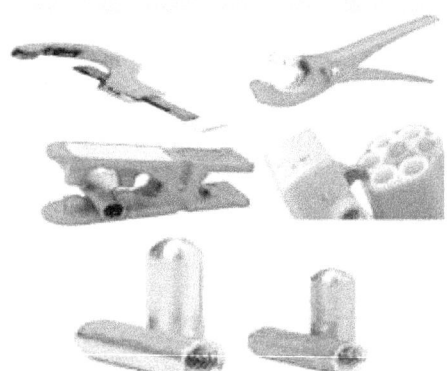

It is essential to always use the prescribed cutting tools to produce a clean and straight cut to a microduct, before inserting a coupler.

Bullet
Fixed to the end of the fibre cable before blowing starts, to guide the cable through the microduct.

Sponge

Used for removing dirt and/or moisture from a microduct.

Shuttle

Used to check the microduct for bends, kinks or blockages.

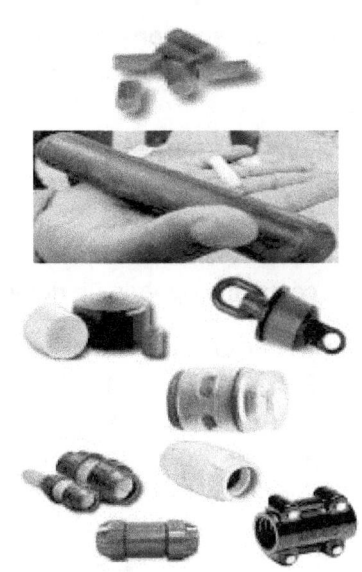

Caps & Plugs

Used for the permanent or temporarily sealing of unused microducts to prevent water and dust from getting in.

Use only the prescribed connectors and end-plugs.

Couplers

Used for the air-tight coupling of microducts.

De-bur Tool

Allows for one to round-off a microduct after cutting - before adding a coupler-this will produce an air-tight connection.

Locking Tool

Used to firmly lock the microduct into a connector.

Cable Carriers

Parachutes capture the air within a duct, tugging along with it the cable. In the absence of air pressure, they collapse allowing for easy pullback. An alternative is to use an Inflatable Carrier.

Aerial Works

Deploying fibre above ground removes the need for digging and particularly useful when the ground is undulating, rocky or both - and has become an increasingly popular option in both urban and rural environments.

Fibre in a duct solutions have a major aesthetic advantage - this because once installed, they are invisible, leaving no mark on the landscape. However, there are a number of reasons for choosing an aerial solution, such as:

- Aerial fibres are typically much faster and cheaper to deploy than buried networks.
- The planned route may be undulating, rocky or both, making digging less appealing.
- All-Dielectric Self Supporting (ADSS) cables can be erected in close proximity to power transmission lines. This of course, allows for pole sharing, which of course, reduces installation costs and speeds-up deployment.

Cable Jacket Colouring

- Polyethylene (PE) is the material of choice for use as an aerial OSP cable jacket.
- The performance of raw PE can degrade rapidly through exposure to Sunlight - Carbon Black absorbs the UV light and subsequently dissipating this energy as harmless heat.
- Jacket colours other than black are used for enhancing identification only.

Pole Handling PPE

1   Safety boots with steel caps.
    Protective clothing with long sleeves.
    Shoulder pads.
    Gloves.
    Hardhat.

Transportation of Poles

1   Poles must never exceed the 0.5m vehicle overhang and must have a red flag secured on the overhanging end.
2   Poles that are loaded onto a pole carrier must be secured to ensure that the cargo does not move while it is in transit.

Pole Off-Loading Procedure

1   Ensure that the removal of any one pole will not cause shifting or rolling of any of the remaining poles.
2   Step 1: Unfasten the poles.
    Step 2: Slide one pole at a time towards the rear end of the vehicle.
    Step 3: When the pole reaches its equilibrium point, the persons on the vehicle must raise their end slowly.
    Step 4: The persons on the ground slowly pull the pole until 1m of it is left on the back of the vehicle bed.
    Step 5: The persons on the ground receive the pole and gently place it on the ground.
3   A pole must never be dropped on the ground, as this could damage the pole and/or cause injury to team members.

Pole Handling Ratios

1   7mpole =4 people
    8m pole =6 people
    9m pole = 8 people or a mechanical aid
    10m pole = mechanical aid.

11m + pole = mechanical aid

Climbing Ladders

1 Keep hands free of tools or materials when climbing or descending a pole or ladder.
2 Workers climbing up or down ladders must always face the ladder and maintain a 3-point contact. This effectively means that 2-hands and 1-foot or 2-feet and 1-hand must be on the ladder at all times.
3 Ladder must be positioned correctly (1-4 ratio).
4 Ladder must be properly secured (lashed and held).
5 Ladder must be in a good condition.
6 The ladder must suit the application.
7 A worker must be correctly positioned on the ladder.
8 A safety harness must be worn and secured to the pole once the working position is reached.
9 Never climb intermediate poles if the span they support is being placed under tension.

Survey

1 Survey rods must be planted in line at selected pole positions so that, when erected, the poles will be in a straight line.
2 A spirit level must be used to verify that there is no lean to the rods.
3 As the survey advances, the rear rods used for lining up - will be withdrawn and survey pegs driven into the ground in the exact position previously occupied by the survey rod.
4 The location of the poles to be erected along roads shall be in accordance with the way leave drawings and conditions stipulated by the authorities concerned.
5 Square wooden pegs shall be used to mark the position of every pole, stay or strut.
6 The numbering (or other details) and marking of the wooden pegs shall be done as agreed upon by both the client and contractor.
7 The tops of pegs that show the positions of angle poles must be marked with blue lumber crayon crosses.
8 A survey peg for a strut position must show the approximate spread of the strut.

Survey Equipment and Tools

1 Digital camera spare batteries.
2 GPS with tracking function and spare batteries.
3 DCP tester.
4 Tape measure.
5 Measuring wheel.
6 Survey pegs and hammer
7 Road cones.
8 Clip board, survey sheet/note book and stationary.
9 Construction vehicle – with signage and an orange light mounted on the rooftop.
10 Road map and/or proposed route detail drawing/s from the client.
11 Reflector jacket.
12 Personal Identification.

Survey - Gather Route Information

1 The information on this route must accurately indicate distances.
2 Take photos of all obstacles on the route (existing services, bridge crossings, rocky areas, buildings, built-up areas, paved/tarred areas, wetlands, overhead obstacles, etc.).
3 Identify all obvious landmarks where the route changes direction (take photos).

4   Note down any road repair work necessary - record distances and GPS coordinates.

5   Provide for a series of DCP test readings along the route and document the exact positions.

6   Description of the topography along the route (sloping, edge of cliff, adjacent to lake, forest surroundings, rivers, swampy areas, etc.) - record distances and GPS coordinates.

7   Description of the ground condition along the route and distances (rocky, sandy, grassy, clay, etc.) - record distances and GPS coordinates.

8   Indicate the distance to the nearest town, where the civil works material (sand, cement, stone, water, tools, etc.) can be sourced from.

9   Locate possible warehouse/camp sites where material can safely be stored.

10 Indicate the availability of hospitals / clinics / police stations along the route - in case required during operational activities.

11 Plan the route to allow for projected road or rail deviations.

12 Double-check recorded details on the return journey.

Pre-Install Meeting

A pre-install meeting or meetings must be held to discuss the survey results, the optimum pulling sites, span lengths, installation equipment and hardware requirements, logistics, splice locations, terrain and other vital installation topics.

Checks to be undertaken prior to commencing with the planned aerial work

1   Does the contractor have approved aerial route drawings, signed by the client?

2   Do the drawings show the alignment of the aerial route within the wayleave specification?

3   Are the wayleaves in place? (copies must be kept on site at all times).

4   Have the locations of existing services been marked and shown on drawings?

5   Are the aerial route drawings being marked indicated on which side of existing road/pathway to stay?

6   Has the accessibility of poles to splicing vehicles been considered?

7   Does the cable have a UV resistant cable jacket?

Wooden pole inspection (prior to planting)

1   Correct type of pole supplied? (length and thickness)

2   Excessively bent or cracked poles should never be used. Horizontal cracks perpendicular to the grain of the wood may weaken the pole. One large knot or several smaller ones at the same height on the pole may be evidence of a weak point on the pole.

3   Inspect the pole for evidence of termites or ants.

4   Ensure that all poles are fitted with 'end plates' and strapping at both ends.

5   The poles should never be off loaded and stacked on the ground for long periods as this could cause damage to the poles as well as the environment.

6   Hammer Test (existing poles): Rap the pole sharply with a hammer weighing about 1kg, starting near the ground line – then continue upwards around the pole to a height of approximately 1.5m. The hammer will produce a clear sound and rebound sharply when striking sound wood. Decayed areas will be indicated by a dull sound or a less pronounced hammer rebound.

Hole-digging Tools

1 The tools provided for hole-digging include picks, shovels, earth augers, crowbars, drills and sledge hammers. The tools to be used for any particular work are determined predominantly by soil conditions.

2 On large projects and wherever ground conditions permit, hydraulically powered Earth Augers can be used. It looks much like a corkscrew and produces extremely clean holes.

Pole Holes

1 All excavations for pole holes will be such that the survey peg indicates the centre of the hole.

2 If the holes are too large, the soil will be unnecessarily disturbed and the poles will not be supported by solid earth. (A diameter of approximately 400mm is recommended).

Length of Poles	Plant Depth
6m and shorter	0.9m
7m to 8m	1.2m
9m and longer	1.5m

3 Where a hole is dug on sloping ground, the depth of the hole shall be measured from the lowest point on the ground surface.

4 In extreme rocky conditions where holes cannot be excavated to the specified depth, an arrangement between contractor and client can be reached for poles to be set in concrete.

Poles set in Concrete

1 Where poles are planted in soil that is difficult to compact, such as sand and swampy areas and in extreme rocky conditions, the poles can be cast in concrete.

2 Only new wooden poles can be set in concrete.

3 The hole must be circular in shape. The hole diameter must be kept to a minimum, but be sufficiently wide to accommodate at least 85mm of concrete between the sides of the pole and the undisturbed ground.

4 The concrete to be used must be made from a mixture of 1 part cement, three parts sand and three parts crushed stone (1:3:3 mix - 15MPa).

5 Concrete must not be compacted around the poles, but thoroughly tamped around the pole with a suitable wooden stick, until the hole is filled.

6 The bottom of the pole must be allowed to "breathe" – therefore, backfill with 10cm of soil before pouring concrete.

Pole Spacing

1 It is advisable to maintain a uniform span length and depart from this only when it is rendered necessary by conditions such as: (1) uneven ground (2) sharp bends (3) or to avoid dangerous positions. This may necessitate the planting of additional poles or omitting of poles.

2 Steel measuring wires for standard span lengths should be made up locally. When the length of span has been chosen the appropriate wire should be used to determine the distance between successive poles. A steel tape measure should be used for checking the length of the measuring wire daily during the survey.

Local ADSS Span Lengths	
Type of route	**(m)**
Short span	83
Medium span	250
Long span	500

Pole Planting Process

1 Ensure that all holes necessary for pole dressing are drilled prior to erection.
2 A pole should be erected by laying it on the ground in such a position that by raising the top section,
3 the base should slide into the hole.
4 Backfilling and ramming must take place in 300mm intervals.
5 Where stones are available they should be used to stiffen the holding.
6 During the backfill and ramming process, always ensure that pole plumbness is maintained.

Suggested Pole Planting Work Practices

1 Avoid dongas, culverts, drains or water channels.
2 Avoid obstructing private roads and entrances.
3 Restrict road crossings to a bare minimum, and if possible, stick to the same side of the road throughout.
4 Avoid trees and where not possible, select a position which will minimise interference from trees – even at the expense of construction costs being increased slightly by this action.
5 Along national and other proclaimed roads the poles and stays should be located in the position agreed to by the Road Authority and as indicated on the wayleave.
6 Keep the route as far away as practically possible from power lines.
7 Where the ground is very soft, poles may be planted 300mm deeper than specified, but only if the necessary vertical clearance is maintained.
8 Ensure that all holes necessary for pole dressing are drilled prior to erection.
9 Maintain a distance of at least 1m from trig beacons and stations.
10 The principle to be followed in all cases is that neither stays nor poles are to be planted where they are likely to cause obstruction or to be dangerous to users of the road, or where they are likely to interfere with ordinary road maintenance such as the clearing and trimming of the edges of the road or the cutting of drains, gutters, etc.
11 In railway reserves, the poles should be located as close as possible to the boundary fence.

Types of Stays

Terminal stays

Provided where the route starts and ends. This stay must be on the side of the pole opposite to the direction of the cable route.

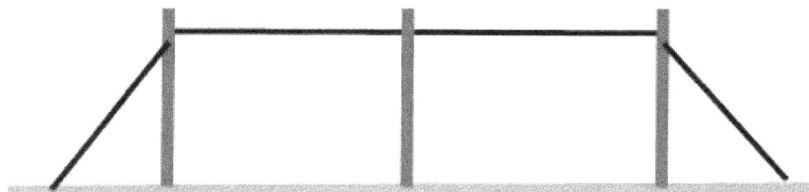

Line stays

Installed at every 13th pole along the route or spaced alternatively as per specification. Line stays must be installed on poles either side of rivers and road crossings where normal span lengths are exceeded.

13th Pole

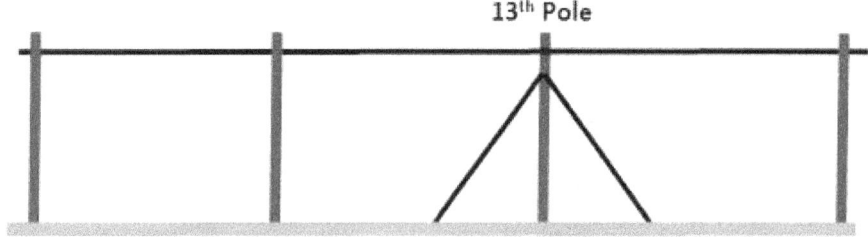

Wind stays & Angle stays

Wind stays are used to stabilize a cable route against wind. Fitted at 90° against the direction of the cable route and on either side of a pole.

Angle stays are used to counter-act a change in direction of the cable route by more than 15° or as per client specs.

Stay Guards

Stay guards must be fitted on all stays exposed to pedestrians, cyclists or vehicles, to make them more visible.

Struts

Struts can be used as an alternative to where stays create traffic hazards, block roads or where a property owner objects to the fitting of a stay.

Struts must be installed on the opposite side of the pole to where the angle stay would have been fitted, to counter-act cable strain.

Stay Holes

1. The cross-section of the hole shall be confined to the smallest size necessary to accommodate a stay plate.
2. The depth of stay holes shall be 1.5 meters or at such a depth where the unthreaded portion of the stay rod protrudes by no more than 25mm above ground level.
3. Stay rods without plates may be used where solid rock is encountered. The stay rod is now inserted in a hole drilled into the rock and secured with cement.
4. In difficult to dig ground conditions shallower holes are allowed subject to approval and shall then be backfilled using concrete.

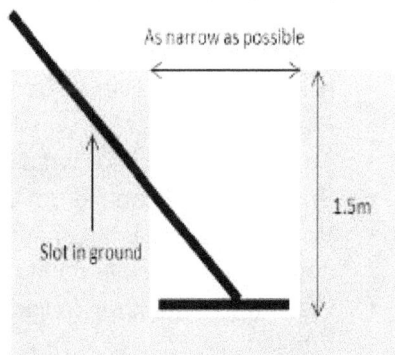

Spread/Height Ratio

1. The spread is the distance from the foot of the pole to the point to where the stay enters the ground.
2. The height is the distance from the ground to the pole attachment.
3. Wind stays shall have a spread/height ratio of 0.6:1
4. Terminal and line stays has a spread/height ratio of 1:1

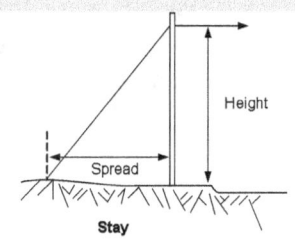

Termination of Stay Wire to Pole or Crosshead

1 Wrap the top pre-formed "make-off" without overlapping around the pole twice at the prescribed height with ends meeting.
2 Twist the top of the preformed "make-off" around the stay wire.
3 Cut the stay wire at the correct length to ensure that the proper spread/height ratio is maintained.
4 Place the bottom of the preformed "make-off" through the crosshead eye.
5 Then pull tight and cut the suspension wire in line with the crosshead and twist the bottom preformed "make-off" around the stay wire (ensure that the crosshead is threaded to the outer most of the stay rod).

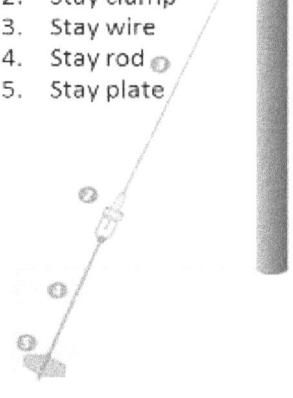

1. Pole bracket
2. Stay clamp
3. Stay wire
4. Stay rod
5. Stay plate

Cable pay-off - over the top or from the bottom?

– Jetting / blowing cable
– Hauling cable
– Figure-eighting cable
– Erecting cable

As for me, over the top makes much more sense. However, advocates of from the bottom, when jetting / blowing cable do exist. This simply reinforces our conviction that ultimately, faced with this choice, a Contractor must determine which option is favoured by the Client.

Pre-Installation Cable Drum Inspection
- Always wear hardhats, gloves and safety boots (with steel toe caps).
- Check that the cable specified, has been procured.
- Inspect the cable drum for signs of excessive weathering and/or damage.
- Drums must be transported or stored with their battens intact.
- Never accept delivery of a cable should the drum is damaged.
- Plastic foil wrap must remain in place until cable placement.
- To remove plastic foil wrap on a cable, do not use sharp tools.
- Ensure that all cable drum bolts are all tightened.
- Verify that nails, bolts or screws on the inside surface of drum flanges are counter-sunk to avoid damage to the cable during placement.
- Place the cable drum in line with the intended direction of deployment, to prevent reel flange-cable rubbing.
- Cable ends must always be sealed - using pre-formed or heat shrinkable end caps.
- Using tape for sealing cable ends is considered unsuitable.
- Always roll the drum following the direction of the arrow.
- Drums must be chocked to prevent them from moving or rolling.

OTDR Pre-Test

- It is vital than an OTDR is used to test all fibres before the installation begins.
- Testing shall be done on all fibres in one direction at 1550nm, using a pulse width of 30ns.
- Traces will be stored and an electronic copy submitted to the client.
- Should a cable be installed without an OTDR pre-test - a supplier can claim that the installer assumed liability upon installation.

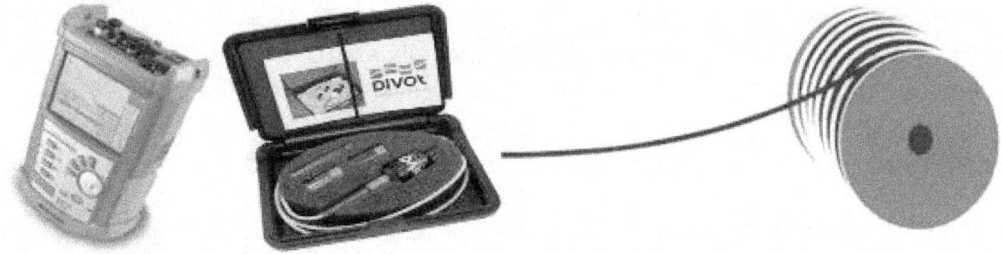

Aerial Cable All-Dielectric Self-Supporting (ADSS) Installation

To be cost-effective aerial deployments require access to existing poles, which may be restricted or need to be strengthened. Self-supporting aerial cables are reinforced with internal strength members to support their own weight.

2 ADSS Installation Methods

1. Installation Process (Conventional Method)

1　To start with, a UV resistant cable jacket is required for all aerial applications.
2　Orientate the drum so that the natural payoff direction faces the pulling direction.
3　To eliminate possible cable contact with the ground, play the cable off from the top of the drum.
4　Fit cable pulley boxes / wheels to every pole on the route, for the length of cable to be erected.

　　　Pulley Wheel　　　　　　　　Pulley Box

5　Feed the pulling rope through the pulleys. It is extremely important that the pulling rope and the ADSS cable have the same diameter.
6　Make a hauling eye at the end of the cable by removing a piece of the cable sheath (250-300mm). Next, the Kevlar of the cable is then wound around the cable and attached to the cable using a 25/8 heat shrink sleeve.
7　Place the drum at least 50m away from the pole where the cable is to go through the first pulley. This will prevent the cable from bending too much while being pulled.
8　Attach a break-away swivel to the end of the hauling rope and then attach the other end of the break-away swivel to the hauling eye of the cable - erecting can now begin.
9　Cable lengths of up to 6000m can be erected with one haul, if the terrain allows for it.
10　Radio Communication between persons at the drum, alongside the cable-end and the hauling team must be maintained.
11　When hauling the cable, a person with a two-way radio must walk alongside the cable-end to ensure that the cable is not twisting with the rope, especially at angle-poles.
12　The hauling team must haul the cable evenly and prevent jerking. The person(s) at the cable drum must "feed" the cable off the drum at the same speed at which the cable is being hauled. There must be no strain on the cable between the drum and the first pulley.
13　A good rule of thumb is to keep the pulling tension to ½ that of the sagging tension (see sagging).
14　When removing a pulling grip, 3-5 meters of adjacent cable must be cut-off and discarded.

2. Installation Process (Figure 8 method)

1　This method should be used only when the terrain is such that the conventional method cannot be used. Place the drum approximately halfway along a long hauling section.
2　Follow steps (1) to (8) as described above (Conventional Method).
3　The one half of the cable length is hauled in the one direction.
4　The balance of the cable is then completely run off the drum into a figure 8 on a tarpaulin, after which it is hauled in the opposite direction.
5　The figure 8 method should not be used for cables longer than 4000m as it becomes risky to manage a coil longer than 2000m, without potentially damaging the cable.

Figure-8-ing Cable

  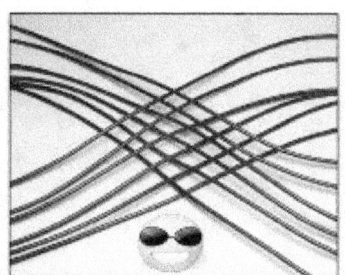 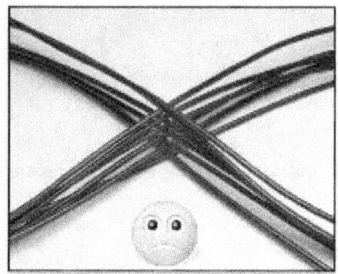

1. A figure-8 puts a ½ twist in on one side of the 8 and takes it out on the other, preventing twists.
2. Set up two traffic cones ± 5m apart – then pay the cable off the top of the reel and loosely weave it around the cones in a figure-8 pattern.
3. In order to pull from the figure-8 - it is necessary to flip it over, a task that requires a minimum of 3-people.

	Ground Clearance	Clearance (m)
1	Non-electrified railways	6.1
2	National road	6.5
3	Abnormal provincial roads	7.5
4	Other provincial roads	6.1
5	Public roads	6.1
7	Private roads or railway tracks to in or near towns	4.9
8	Country roads or railway lines, or over veiled or private lands other than (10)	3.7
9	Cultivated farmlands and across points of entry in cultivated lands	4.9

Local Thimble Type Dead-End Sizes

Cable Dia. (mm)	Colour Code
9.2 - 9.5	Black
10.4 - 10.8	Red
11.5 - 12.2	White

In order not to damage the cable jacket, it is immensely important to use the correct size dead-end, NB, colour codes differ from supplier-to-supplier.

ADSS Terminations and Support Types

1. Where an aerial cable ends (i.e. goes underground or into a splice joint), it is a True Termination. If the span continues (i.e. the cable is not cut) and is called a False Termination.
2. Where do we find False Terminations? On either side of road crossings | Every 500m (or as per client spec) | ≤ 10° angles (or as per client spec).
3. At False Terminations, a 100mm loop (goose neck, drip loop) in the cable, must be left at the pole between the two dead-ends.
4. At an angled pole, the cable must always pass on the front of the pole - never behind.
5. To support the cable at intermediate poles, attach a tangent support or a support clamp to the S-hook.
6. After a cable is terminated, wait for a while (± 10min) to allow the cable to stabilise, before securing the cable in the tangent support or support clamp, for intermediate support.

| Tangent Support | False Termination | True Termination |
| Support Clamp | False Termination | |

Securing of Cable to Poles (Termination)

1 Break-away swivels are used alone or in conjunction with dynamometers to ensure that the maximum pulling tension is not exceeded.

2 To terminate the cable at a terminal pole (beginning or end of route), a preformed dead-end, is wrapped around the cable and hooked onto a single S-hook.

3 With several sizes available, ensure that the correct size dead-end is used.

4 The (17m or so, on each cable) splicing slack is then coiled in a 500mm coil and secured to the pole as high as possible from the ground. Start coiling by rolling the slack cable like a wheel. This will ensure that no twists are put in the slack, which could result in the fibres being damaged.

5 When a route deviates with an angle ≤ 10°, the cable must be terminated as explained below:
   a. Fit a double S-hook on the angle-pole.
   b. Wrap a temporary dead-end around the cable beyond the angle-pole.
   c. A number of workers must then pull on the rope until the desired tension is obtained.
   d. While the tension is held steady, a person on top of the pole wraps a dead-end around the cable and hooks it onto the S-hook.
   e. Remove the temporary dead-end from cable.
   f. A dead-end is then wrapped around the cable in the opposite direction and hooked onto the termination hook.
   g. As mentioned, a 100mm diameter loop (goose neck, drip loop) in the cable, must be left at the pole between the two dead-ends.

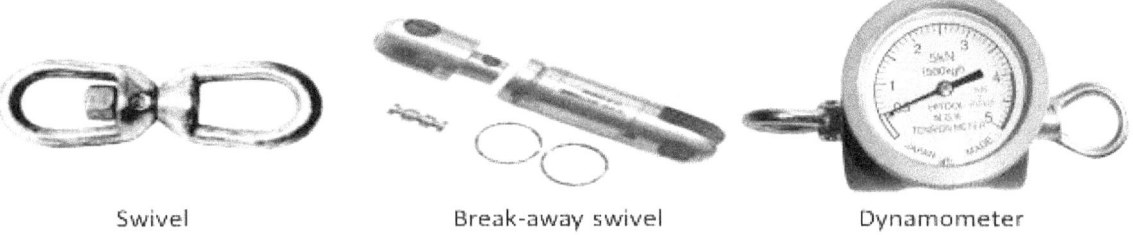

| Swivel | Break-away swivel | Dynamometer |

ADSS Sagging- Line of Sight Method

1 The installation sag is typically 1% of a span length. Less sag will require stronger cables.

2 To my knowledge, the 'line of sight' method is our best option.

The approach taken here requires for the exact distance between poles to be established. 50m as shown in the example below. In essence, the far-side pole is marked using bright coloured tape with the appropriate mid-span sag distance from the attachment height i.e. .5m (1% of 50m).

Next, the installer returns to the first pole and places his/her line of sight at that same height as the marker tape on the second pole, to line-up the mid-span sag distance.

This person must have radio contact with the team tensioning the cable and give instructions of how much to tighten the cable, for the belly of the sag to rise and match the coloured tape mark on the opposite pole.

One or two spans between dead-end locations should be checked using this method.

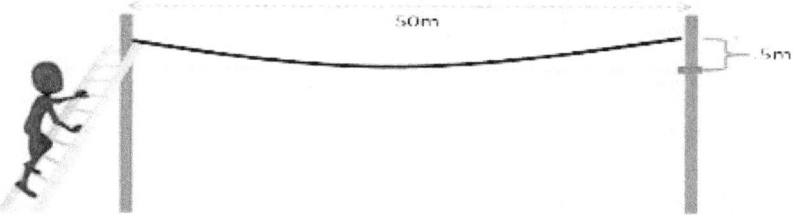

3   If the sagging operation takes place from the pulling end back to the payoff end, then the maximum amount of cable can be recovered.

4   When removing the pulling grip, 3-5 meters of adjacent cable must always be cut-off and discarded.

Cable ratchet hoist

1   A ratchet hoist is used to tension the erected cable.

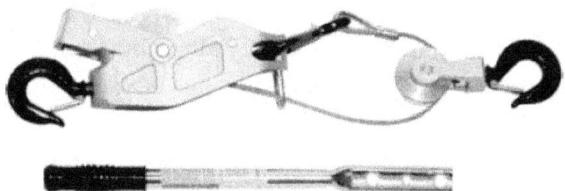

Cable Slack

1   17m (on each cable) of slack is recommended, 2m for splicing and 15m to get down the pole into the splicing vehicle.

2   The slack must be coiled before the joint is secured to the pole/arm. Start the slack coiling process with the joint as the leading end (rolled like a wheel). This will ensure that no twists are imparted to the cable.

3   On completion of the coiling, place the coil into the slack box or onto the slack cross and secure the joint closure at a suitable position on the arm/pole.

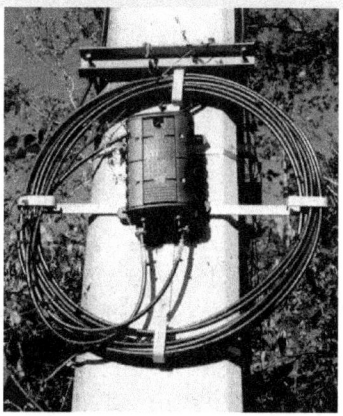

Anti-vibration damper

1   To reduce the natural vibration of the cable, the damper is helically sized to grip the cable on one end and to interact mechanically on the other end.

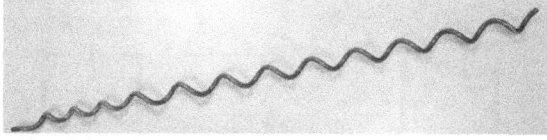

Damage by veld fire to aerial-cables

1   Once installed, aerial cables are impacted by the conditions of the environment they are in.

2   A fact often overlooked, is that grass or bush under aerial-cable routes create a fire hazard when dry. Such grass or bush must be cut before it dries during autumn.

Cable Hauling

If we can rewind a bit, the advantages of using air-assisted installation techniques over cable hauling include; longer installation distances possible, less force is exerted on the cable, etc. However and sure enough, all too often we are still required to haul cable. And, if so, Heavy Duty Duct is the recommended pull-able cable used for hauling.

Pre-Installation Check

1   Prior to any hauling, the following must be confirmed:
   a. Maximum allowable pulling tension
   b. Minimum allowable bending radius during installation
   c. Minimum allowable bending radius after installation
   d. Cable length
   e. Cable length required at the splicing locations

Duct Rodding

1   Rodding is used to clear the duct passage and install the pulling rope, using fibreglass pushrods. An alternative is to use air-blowing device.

2   a. Push the rod into the duct until the front end of the rod reaches the adjacent HH / MH.
   b. Attach the pulling rope to the end of the rod at the adjacent HH / MH.
   c. Next, pull the rod and pulling rope back through the duct
3   If the air pressure method is preferred, the pulling rope will be blown through the duct until it reaches the adjacent HH / MH.

Duct Testing and Cleaning

1   To clean the duct, attach a cylindrical brush or close-fitting mandrel and a second rope to one end of the installed pull rope and pull this through the duct.
2   If the cable sample or mandrel cannot pass through the duct, consult with the Client on whether to switch to an alternative duct or to repair the designated duct.

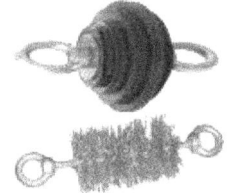

Centre-Pulls and Back-feeding

One is often more hopeful than optimistic to install a whole drum of cable in one operation, when faced with hauling a long length of cable. Lubrication helps but more than often, does not completely resolve this issue. To assist with alleviating this issue, specialized techniques, such as Centre-Pulls and/or Back-feeding, can be used and applied with either the manual pull or winch methods.

Centre-Pulls

1   Communication, directed by a team leader with team members positioned at each HH / MH is essential, for that the pulling action to be achieved in a synchronized manner. Communication via the use of either a walkie-talkie or two-way radio is desirable.

2   In a centre-pull operation, set up the cable reel near the centre of the duct run to be pulled.
3   To avoid cable rubbing against drum flanges keep the drum level - orientate the drum so that natural payoff direction is towards the pulling direction. Always use cable jacks or a cable trailer to lift the drum during the hauling process.
4   Align the cable drum so that the cable can be routed from the top of the reel into the duct in as straight

a path as possible.

5   Make use of a flexible cable guide – placed between the HH / MH lid and duct to be used.
6   Ensure that there is a swivel between the cable sock and the hauling rope.
7   Pull the cable in one direction to the intended HH / MH.
8   Uncoil the remaining cable in a figure-eight configuration.
9   Flip-over the figure-eight so that the pulling-eye end of the cable is on top. As mentioned previously, this can be accomplished by 3-installers, one at each end of the eight, and one at the centre.
10 Pull the exposed end of the cable in the opposite direction to complete the pull. Hand-feeding of the cable paying off from the figure-eight is required.
10 A warning marker (coloured tape or similar material) may be attached to the pull-line at ± 10m in front of the pulling grip to alert observers at HHs / MHs that the cable is approaching.
11 Place an end cap on all bare cable ends in HHs / MHs, to prevent moisture and/or dirt intrusion.

Back-feeding

1   In a nutshell, back-feeding is used to provide a series of shorter, lower-tension pulls in one direction – utilising figure-eighting, where necessary.

Pulling Tension

1   The maximum allowable pulling tension on fibre cable can vary and is dependent on the cable construction. The maximum tension for a particular cable can be found on the cable spec sheet.
2   Except for short runs or hand-pulls, tension must be monitored.
3   A dynamometer is to be used to monitor tension in the pull-line
4   The use of a breakaway swivel can be used to ensure that the maximum tension of the cable is not exceeded and should be used as a fail-safe rather than a primary means of monitoring tension.
5   Cable lubricant is recommended for most fibre optic cable pulls as a means of lowering pulling tension.
6   Pulling a new fibre optic cable over an existing one is never recommended due to the possibility of entanglement.

Pulling Grips

1   Pulling grips provide effective coupling of pulling loads to Kevlar strength members or cable jackets.
2   On cables with no Kevlar strength members, a wire mesh pulling grip and swivel can be used during cable pulls.
3   A swivel must always be fitted between the hauling rope and cable sock to prevent the cable from being twisted by the hauling rope.
4   When removing a pulling grip, 3-5 meters of adjacent cable must be cut-off and discarded.

Hauling Rope

1   An 8mm polyester or polyaramid pull rope must to be used for optical fibre cables that are to be pulled in by hand.
2   When using a hauling winch, a 12,5mm or thicker polyester or polyaramid rope must be used as hauling rope.
3   Ski-ropes manufactured from nylon are not suitable, as their stretch factor is too high and they can potentially cut into ducts.
4   Steel hauling rope or galvanised wire must never be used as they can damage ducts and/or cables.

Duct Fill Ratios for hauling

1   Fill ratios are calculated by comparing the area of an inner diameter cross-section of the inner-duct to the outer diameter cross-section area of the fibre optic cable. For optimum hauling performance, it is recommended that the cable-to-duct diameter fill ratio does not exceed 65%, or as per cable spec sheet.

Duct Fill Ratio

$$\frac{\text{Cable Diameter}}{\text{Inner Duct Diameter}} < 65\%$$

Access Build

By now, we are familiar with the popular and conventional installation practices (which do apply here) - however, each last drop (the final connection to a home or building) is unique. Traditionally fraught with deployment challenges to be overcome and the usual suspects are:
- Multiple tight bends
- Space constraints

This of course is why the last bit of an installation is normally the most expensive and time consuming part of any overall fibre project. This pickle is traditionally met with an echoing yawn from customers and gnashing of teeth from contractors.

Bend-insensitive Fibre

It has been long understood that an inherent limitation of optical fibre is macro bend attenuation. Put simply, installers face increased loss with tight bends. The focus of this limitation has been amplified as optical networks continue to move closer to where people work and live.

Bend-insensitive SM fibre (ITU-T G.657A) has an innovative design (a lowered refractive index in the cladding area) that enables it to significantly reduce macro bend loss even in the most challenging bend scenarios.

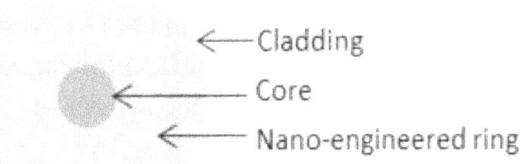
←—Cladding
←——— Core
←——— Nano-engineered ring

SM fibres complying with ITU-T G.657A was developed with the purpose of use at FTTH sites. G.657A category fibres are compatible with the standard G.652D fibre, it was designed to replace.

Indoor Fire Code Ratings

Every cable installed indoors must meet fire codes. That means the jacket must be rated for fire resistance. Most indoor cables us PVC (polyvinyl chloride) jacketing for fire retardance. All premises cables must carry identification and flammability ratings as per the **NEC (National Electrical Code) paragraph 770.50.**

Experience with fire accidents of buildings reveal that most of the casualties happen due to the smoke generated. Smoke is considered to be root cause of deaths and therefore international guidelines for cabling of multi dwelling apartments, and office buildings must be followed.

OFNP stands for Optical Fiber Non-conductive Plenum. OFNP cables have fire-resistance and low smoke production characteristics. They can be installed in ducts, plenums and other spaces used for building airflow. This is the highest fire rating fiber cable and no other cable types can be used as substitutes.

Outdoor cables are not fire-rated and can only be used up to 15m indoors.

Cables without markings should never be installed as they will not pass inspections!

It is not simply the technology, but also a human consideration required when specifying a particular optical fiber cable for indoor use.

The Deployment of Optical Fibre Infrastructure on Dolomitic Land

Dolomite bedrock are notorious for the occurrence of sinkholes. The deployment optical fibre infrastructure involves the installation of sub-surface ducts, HHs and / or MHs. On dolomite land, the installation must comply with the requirements of SANS 1936-3 for dry services. These installations differ from the typical projects envisaged by the code writers due to the widespread nature of the work and the low impact that such services have on dolomitic stability if correctly installed and managed.

Appointment of a company competent person:
This person must be a registered Professional Engineer or Professional Technologist who:
   a) Complies with the requirements for a competent person given in clause 3.3 b) and c) of SANS 1936-1:2012
   b) Has achieved Competence Level 3 (Experienced Professional) given in Figure A.1, Annex A, SANS 1936-1:2012

The Risk
With a dry engineering service, the biggest risk to be considered is the possibility that the service may become a conduit through which water can be introduced into the ground. The most likely points where water can gain access to the system and potentially cause sinkholes are:

   a) Through non-water tight HHs / MHs.
   b) At the point where the duct enters the HHs / MHs.
   c) By interception of leakage from other services.
   d) By infiltration of water from the ground surface through the backfill of both trenches and HHs / MHs.

Mitigating Risk
   a) Ensuring that the HHs / MHs themselves, their covers and the connection of the ducts into the HH / MH walls have an acceptable degree of water tightness.
   b) Ensuring that existing water bearing services are not damaged during the laying operation and that any existing leakage from such services is reported to the relevant authorities.
   c) Ensuring that the backfill to service trenches and HHs / MHs comprises of the same excavated material compacted to a higher density than that of the surrounding ground thereby reducing its permeability.
   d) By ensuring that the ground surface above the trench is finished level with the surrounding areas to prevent surface drainage being impeded.

Distribution of Dolomite in SA

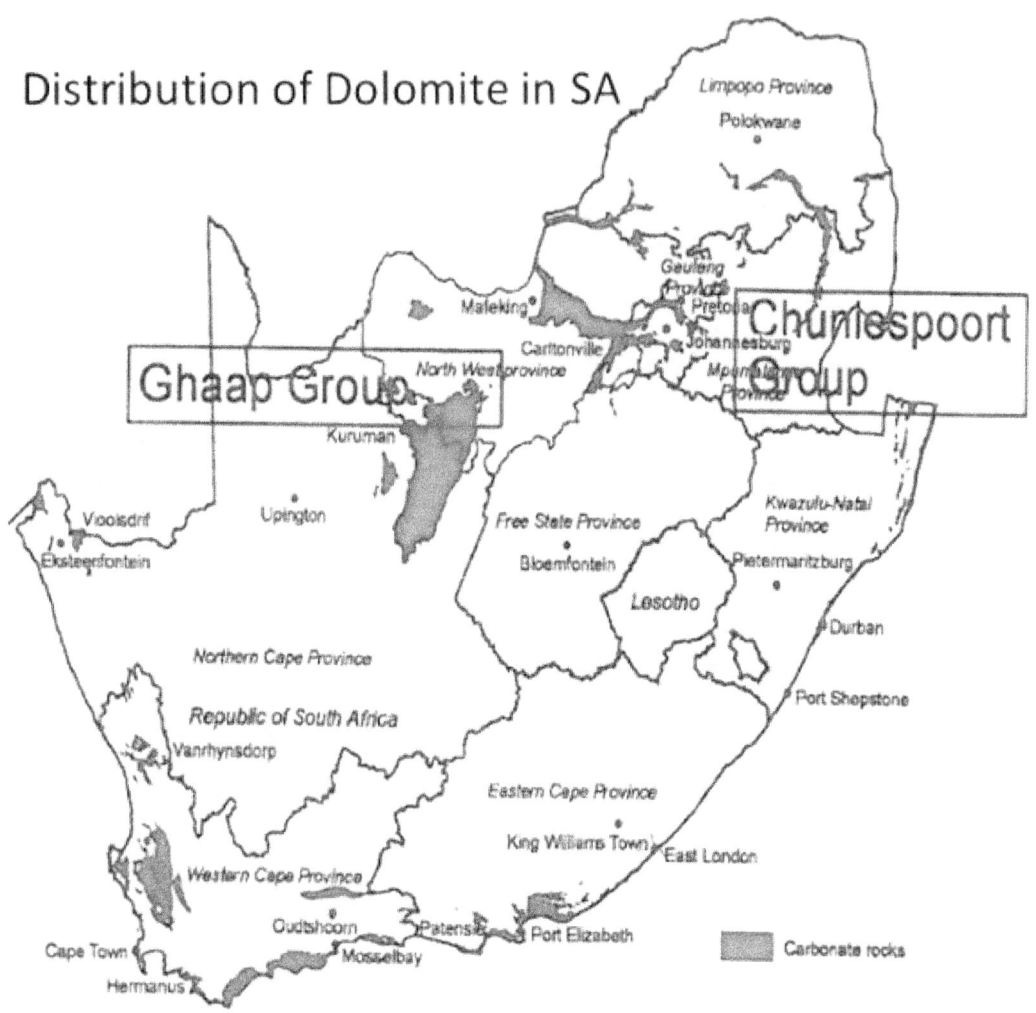

Quality Management Plan (QMP)

This brief QMP Outline is provided for informational purposes only. A QMP describes the overall policies, program, responsibilities, procedures, and the means of ensuring that all executed work will be in conformance with the relevant client specification/s. Without it, one is unlikely to avoid the feeling of being a lobster in the pot of water that has gone beyond warm.

Quality Control (QC) versus Quality Assurance (QA)

You can think of QC as the activities that are used to **EVALUATE** whether or not your product or service meets the specified quality requirements.

QA, on the other hand, is all about **ENSURING** that the product is produced in the right way. It is proactive and concerned about the processes and activities used.

QC versus QA *Joe's interpretation...*

Think of QC as something such as doing periodic checks to see if the pool is sparkling clean, whereas QA aim to make certain that the pool owner does the right things; maintain the pH level, scoop-out leaves, clean-out the strainer basket, brush the walls and tiles, run the pool pump long-enough, backwash regularly, etc.

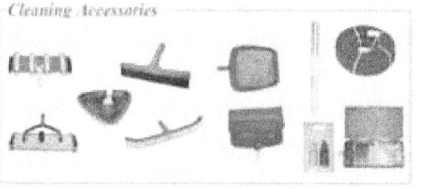

Cleaning Accessories

At its simplest, QC is testing or checking-out a service or product, to make sure that it's OK. The intent is to spot anything that is not-OK, and then to immediately fix it.

QA is aimed at delving into how not-OK can be eliminated.

Quality Control Officer (QCO)
1. The QCO is appointed to be responsible for the implementation of a contractor's QMP.
2. He or she is also responsible for advising and directing site personnel in order for them to understand and carry out their responsibilities diligently.
3. Good communication is critical to the success of a project.

The QMP describes and defines:
1. Participant roles and responsibilities.
2. QC controls to be applied.
3. QC documentation and specifications to be utilised.

A QMP customarily contains 4-phases:
1. Preparatory phase before construction.
2. Commencement of construction.
3. Inspection and testing to deliver untroubled quality compliance and workmanship.
4. Final acceptance of work.

The responsibilities of personnel who manage quality, routinely include:
1. Ensuring that all staff has undergone the appropriate training and certification for the types of construction activities they will be performing.
2. Perform QC inspections of on-going construction work for the duration of the project.
3. Identify, evaluate, and document quality problems.
4. Initiate action to prevent the occurrence of non-conforming work.
5. Recommend or initiate quality improvement solutions.
6. Stop the work when non-conforming work is identified, until the deficiency is corrected.
7. Maintain a Non-Conformance Report (NCR) log.

Training

1. All personnel on a project, will be made aware of the quality requirements to their position.
2. Personnel will be trained to ensure that they possess the necessary skills and knowledge to execute their work.
3. At the start of their job on a project, all employees will receive an orientation on their individual roles and responsibilities.

Goals, Objectives and Performance Monitoring

Progress against established project targets will be reported at weekly team meetings and will include:

1. Are Health and Safety conditions being maintained?
2. Are environmental standards being maintained?
3. Are cost objectives and targets being met?
4. Are quality standards being maintained?
5. Are progress targets being met?

Review and Reporting

1. The project team will hold weekly meetings, to review performance to date and to plan in detail the remaining activities.
2. Minutes will be taken and actions required recorded.

Equipment Inspection and Maintenance

1. Site staff will maintain maintenance schedules of the equipment on site and calibration records will be maintained.

Document Control

1. As-build documents must be kept current.
2. Hard copies of quality files will be kept in agreed locations and will be readily identifiable.
3. The PM will ensure that quality handover documents are approved and forwarded to the client in accordance with an agreed handover process.

Non-Conformance

1. All instances of non-compliant actions, damage or non-conforming products shall be reported to the PM.
2. The PM will detail the nature and cause of the non-compliance, the action (agreed upon) to rectify the issue and any further actions that are proposed to be taken to prevent a recurrence.

Skilled Labour

Some service providers may have their own labour while others rely exclusively on contractors. The trouble is that skilled labour is not always readily available. Because performance is so clearly measurable and apparent, FOA trained labour remains an important requirement, particularly for guaranteeing that a good end product is delivered. Even with the much more resilient and robust bend-insensitive varieties, optical fibre is still glass and requires both skill and knowledge to install.

In what follows, find a few quality inspection sheet examples:

Trenching and Duct Placement		OK	Not-OK	N/A
1	Wayleave agreements			
2	Traffic management			
3	Personal protective equipment			
4	Barricading			
5	Environmental considerations			
6	Businesses / property owner notifications			
7	Location of services			
8	Pilot holes			
9	Power Cable separation			
10	Trench depth			
11	Trench width			
12	Bedding and padding			
13	Vertical or horizontal duct de-coiler used			
14	No duct directional changes			
15	End caps on ducts			
16	Warning / marking tape placed			
17	Backfill OMC			
18	DCP test			
19	Site tidy and clean			
20	Reinstatement			

Manholes and Handholes		OK	Not-OK	N/A
1	MH / HH cover seal			
2	Approved foot-roadway frame and cover			
3	MH / HH walls watertight			
4	Ducts secured			
5	End caps fitted on empty ducts			
6	Ducts sealed in-between wall			
7	Used ducts sealed between cable and duct			
8	Cable slack neatly stored			
9	Cables and ducts labelled			
10	HH/MH marked on the coping			
11	Splice closures secured			
12	HH/MH tidy, clean and dry			

Air-Assisted Installations		OK	Not-OK	N/A
1	Personal protective equipment			
2	Barricading			
3	Two-way communication			
4	DIT results available			
5	Cable under no tension			
6	Cable clean			
7	Bullet on cable end			
8	Designated duct used			
9	Cable-to-duct diameter fill ratio			
10	Transportation duct correct size			
11	Blowing equipment good working condition			
12	Monitor blowing speed and torque			
13	Correct lubricant used			
14	Cable drum correctly positioned			
15	Figure-8 laid on a ground sheet			
16	Good housekeeping			

Cable Hauling		OK	Not-OK	N/A
1	Personal protective equipment			
2	Barricading			
3	Two-way communication			
4	Duct testing and cleaning			
5	Fibreglass pushrods used for rodding			
6	Correct pull rope used			
7	Break-away swivel used			
8	Cable drum correctly positioned			
9	Pulling grip size			
10	Allowable bending radius maintained			
11	Correct lubricant used			
12	Figure-8 laid on a ground sheet			
13	Dynamometer used to monitor tension			
14	Duct fill ratio			
15	17m (on each cable) of slack			
16	Good housekeeping			

Aerial Works Poles and Stays		OK	Not-OK	N/A
1	Personal protective equipment			
2	Transportation of poles			
3	Pole off-loading procedure			
4	Correct type / size of pole supplied			
5	Hole-digging tools			
6	Pole hole depth and width			
7	Uniform pole spacing			
8	Pole dressing holes drilled prior to erection			
9	Backfilling and ramming			
10	Pole plumbness			
11	Correct type of stay fitted			
12	Correct type of strut fitted			
13	Depth of stay hole			
14	Stay spread / height ratio			
15	Stay guards			
16	Ladder/s correctly positioned			
17	Ladder/s lashed to the pole			
18	Correct type of ladder/s used			
19	Good housekeeping			

Aerial Works ADSS Installation		OK	Not-OK	N/A
1	Personal protective equipment			
2	Cable drum correctly positioned			
3	Play the cable off from the top of the drum			
4	Cable pulley boxes / wheels used			
5	Correct pulling rope used			
6	Break-away swivel used			
7	Two-way communication			
8	Pulling grip size			
9	Thimble type dead-end size			
10	Tangent support / support clamp size			
11	Pulling tension			
12	Sagging tension			
13	Allowable bending radius maintained			
14	Ratchet hoist is used			
15	Ground Clearance			
16	17m (on each cable) of slack			
17	Ladder/s correctly positioned			
18	Ladder/s lashed to the pole			
19	Correct type of ladder/s used			
20	Good housekeeping			

Single-mode fibre selection in a nutshell

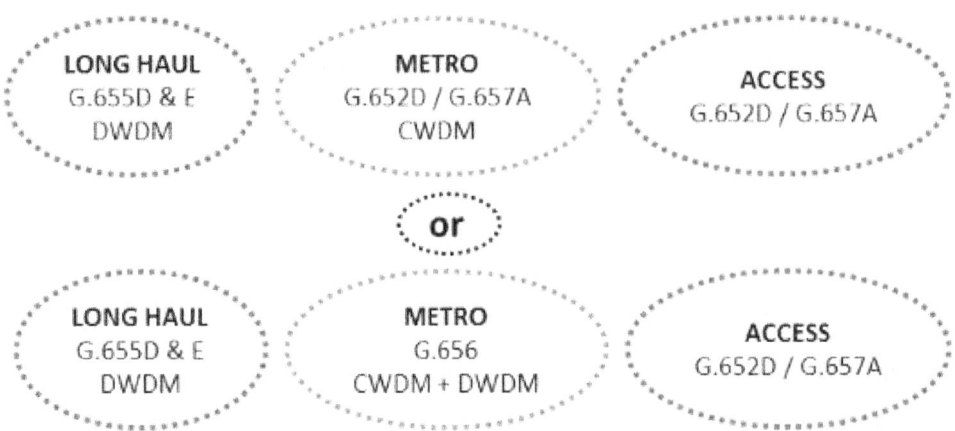

Looking for a single-mode (SM) fibre to light-up your multi-terabit per second system? Probably not, but let's say you were - the smart money is on your well-intended fibre sales rep instinctively flogging you ITU-T G.652D fibre. Commonly referred to as standard SM fibre and also known as Non-Dispersion-Shifted Fibre (NDSF) - the oldest and most widely deployed fibre. Not always a great choice, right? You bet. So for now, let's resist the notion that you can do whatever-you-want using standard SM fibre. A variety of SM optical fibres with carefully optimised characteristics are available commercially: ITU-T G.652, 653, 654, 655, 656 or 657 compliant.

Designs of SM fibre have evolved over the decades and present-day options would have us deploy G.652D, G.657A, G.655 or G.656 compliant fibres.

Note that G.657A is essentially a slightly more expressive version of G.652D, with a considerably better bending loss performance, designed to replace G.652D as a FTTx last drop option. G.657A, contains a ring with a lowered refractive index (less dense) in the cladding and a reduced the mode field diameter - making it bend-insensitive. G.657A is G.652.D compliant in all bands 1260 -1625nm and therefore, treated as one in this discussion.

Dispersion Shifted Fibre (DSF) in accordance with G.653 has no chromatic dispersion at 1550 nm. However, they are limited to single-wavelength operation due to non-linear four-wave mixing. G.654 compliant fibres were developed specifically for undersea un-regenerated systems and since our focus is directed toward terrestrial applications - let's leave it at that.

In the above context, the plan is to briefly weigh-up G.652D, G.655 and G.656 compliant fibres against three parameters we calculate (before installation) and measure (after installation). I must just point-out that the fibre coefficients used are what one would expect from the not too shabby brands available today.

Attenuation

G.652/7 compliant		G.655 compliant		G.656 compliant	
λ	ATTN	λ	ATTN	λ	ATTN
nm	dB/km	nm	dB/km	nm	dB/km
1310	0.30	1310		1310	
1550	0.20	1550	0.18	1550	0.20
1625	0.23	1625	0.20	1625	0.22

Attenuation is the reduction or loss of optical power as light travels through an optical fibre and is measured in decibels per kilometer (dB/km). G.652/7 offers respectable attenuation coefficients, when compared with

G.655 and G.656. It should be remembered, however, that even a meagre 0.01 dB/km attenuation improvement would reduce a 100 km loss budget by a full dB - but let's not quibble. No attenuation coefficients for G.655 and G.656 at 1310? It was not, as you may immediately assume, an oversight. Both G.655 and G.656 are optimized to support long-haul systems and therefore could not care less about running at 1310 nm. A cut-off wavelength is the minimum wavelength at which a particular fibre will support SM transmission. At ≤ 1260 nm, G.652/7 has the lowest cut-off wavelength - with the cut-off wavelengths for G.655 and G.656 sitting at≤ 1480 nm and ≤1450 respectively - which explains why we have no attenuation coefficient for them at 1310 nm.

PMD

G.652/7 compliant	G.655 compliant	G.656 compliant
PMD	PMD	PMD
ps / √km	ps / √km	ps / √km
≤ 0.06	≤ 0.04	≤ 0.04

Polarization-mode dispersion (PMD) is an effect caused by asymmetrical properties in an optical fibre that spreads the optical pulse of a signal.

Slight asymmetry in an optical fibre causes the polarized modes of the light pulse to travel at marginally different speeds, distorting the signal and is reported in ps / √km, or "ps per root km". Oddly enough, G.652/7 and co all possess decent-looking PMD coefficients. Now then, popping a 40-Gbps laser onto my fibre up against an ultra-low 0.04 ps / √km, my calculator reveals that the PMD coefficient admissible fibre length is 3,900 km and even at 0.1 ps / √km, a distance of 625 km is achievable.

So far so good? But wait, there's more. PMD is particularly troublesome for both high data-rate-per-channel and high wavelength channel count systems, largely because of its random nature.Fibre manufacturer's PMD specifications are accurate for the fibre itself, but do not incorporate PMD incurred as a result of installation, which in many cases can be many orders of magnitude larger. It is hardly surprising that questionable installation practices are likely to cause imperfect fibre symmetry - the obvious implications are incomprehensible data streams and mental anguish. Moreover, PMD unlike chromatic dispersion (to be discussed next) is also affected by environmental conditions, making it unpredictable and extremely difficult to find ways to undo or offset its effect.

CD

652/7 compliant		G.655 compliant		G.656 compliant	
λ	CD	λ	CD	λ	CD
nm	ps/(nm·km)	nm	ps/(nm·km)	nm	ps/(nm·km)
1550	≤ 18	1550	2~6	1550	6.0~10
1625	≤ 22	1625	8~11	1625	8.0~13

Called chromatic dispersion to emphasise its wavelength-dependent nature has zip-zero to do with the loss of light. It occurs because different wavelengths of light travel at different speeds.

Thus, when the allowable CD is exceeded - light pulses representing a bit-stream will be rendered illegible. It is expressed in ps/ (nm·km). At 2.5-Gbps CD is not an issue - however, lower data rates are seldom desirable. But at 10-Gbps, it is a big issue and the issue gets even bigger at 40-Gbps.

What's troubling is G.652/7's high CD coefficient - which one glumly has to concede, is very poor next to the competition. G.655 and G.656, variants of non-zero dispersion-shifted fibre (NZ-DSF), comprehensively address G.652/7's shortcomings. It should be noted that nowadays some optical fibre manufacturers don't bother with distinguishing between G.655 and G.656 - referring to their offerings as G.655/6 compliant.

On the face of it, one might suggest that the answer to our CD problem is to send light along an optical fibre at a wavelength where the CD is zero (i.e. G.653). The result? It turns out that this approach creates more problems than it is likely to solve - by unacceptably amplifying non-linear four-wave mixing and limiting the fibre to single-wavelength operation - in other words, no DWDM. That, in fact, is why CD should not be completely lampooned. Research revealed that the fibre-friendly CD value lies in the range of 6-11 ps/nm·km. Therefore, and particularly for high-capacity transport, the best-suited fibre is one in which dispersion is kept within a tight range, being neither too high nor too low.

NZ-DSFs are available in both positive (+D) and negative (-D) varieties. Using NZ-DSF -D, a reverse behavior of the velocity per wavelength is created and therefore, the effect of +CD can be cancelled out. I almost forgot to mention, by the way, that short wavelengths travel faster than long ones with +CD and longer wavelengths travel faster than short ones with -CD. New sophisticated modulation techniques such as dual-polarized quadrature phase-shift keying (DP-QPSK) using coherent detection, yields high quality CD compensation. However, because of the added signal processing time (versus simple on-off keying) they require, this can potentially be a poor choice from a latency perspective.

WDM multiplies capacity

The use of Dense Wavelength Division Multiplexing (DWDM) technology and 40-Gbps (and higher) transmission rates can push the information-carrying capacity of a single fibre to well over a terabit per second. One example is EASSy's (a 4-fibre submarine cable serving sub-Saharan Africa) 4.72-Tbps capacity. Now then, should my maffs prove to be correct, 118 x 40-Gbps lasers (popped onto only 4-fibres!) should give us an aggregate capacity of 4.72-Tbps?

Coarse Wavelength Division Multiplexing (CWDM) is a WDM technology that uses 4, 8, 16 or 18 wavelengths for transmission. CWDM is an economically sensible option, often used for short-haul applications on G.652/7, where signal amplification is not necessary. CWDMs large 20 nm channel spacing allows for the use of cheaper, less powerful lasers that do not require cooling.

One of the most important considerations in the fibre selection process is the fact that optical signals may need to be amplified along a route. Thanks in no small part (get the picture?) to CWDM's large channel spacing - typically spanning over several spectral bands (1270 nm to 1610 nm) - its signals cannot be amplified using Erbium Doped-Fibre Amplifiers (EDFAs). You see, EDFAs run only in the C and L bands (1520 nm to 1625 nm). Whereas CWDM breaks the optical spectrum up into large chunks - by contrast, DWDM slices it up finely, cramming 4, 8, 16, 40, 80, or 160 wavelengths (on 2-fibres) into only the C- and L-bands (1520nm to 1625nm) - perfect for the use of EDFAs. Each wavelength can without any obvious effort support a 40-Gbps laser and on top of this, 100-Gbps lasers are chomping at the bit to go mainstream.

Making the right choice

On the whole, it is hard not to conclude that the only thing that genuinely separates fibre types for high-bit-rate systems is CD. The 3-things, the only ones that I can think of - that is good about G.652D - is that it is affordable and along with G.657A, cool for CWDM and perfect for short-haul environments. Top of the to-do lists of infrastructure providers pushing the boundaries of DWDM enabled ultra-high-capacity transport over short, long or ultra-long-haul networks - needless to say, will be to source G.655/6 compliant fibres.

The End

The Fiber Optic Association, Inc.

1119 S. Mission Road #355, Fallbrook, CA 92028
Tel 1-760-451-3655 Fax 1-781-207-2421
Email: info@thefoa.org   http://www.thefoa.org